How I Shot My Bears

HOW

TWO YEARS'

I

TENT LIFE

SHOT

IN

MY

KULLU & LAHOUL

BEARS

Mrs. R.H. TYACKE

THARAH KARDU

THARAH KARDU
#271/1, Dhamserh
Kullu – 175126
HP, India

tharahkardu.in

How I Shot My Bears: Two Years' Tent Life in Kullu and Lahoul
by Mrs. R. H. Tyacke (born Isabella Edwards)

First published in 1893

ISBN: 9788193875841 (paperback)
ISBN: 9788193875872 (ebook)

BRIEF CONTENTS

CONTENTS

ILLUSTRATIONS

PART-ONE

OUR WINTER IN KULLU

CHAPTER I

AT a time when women have been busy exploring Central Africa and the North Pole, it needs some courage to put on paper a record of wanderings and doings, the only merit of which lies in the fact that their scene was somewhat off the beaten track. Fellow-sportsmen, however, who have tasted the joys of Indian shikar, will fully sympathize with me in my desire to recall to them such delights, as well as to place before the less fortunate some pictures of the lights and shades of sport and camp-life in the Central Himalayas.

Various considerations led us to pitch upon Kullu and Lahoul as a "fresh field and pasture new." We had well-worked that Switzerland, that preserve of India, the delight at once of the traveller, the artist, and the sportsman—Kashmir; we had sought both sport and a genial climate by wintering among the snipe and woodcock in Albania, and with fair results. Now we turned our eyes towards Kullu, as a country little disturbed by sportsmen from India, and where we could combine a not too rigorous climate, fair shooting, and some chance of obtaining the necessaries of life—a consideration indeed, when the nearest railway is two hundred miles distant.

Kullu is situated in 32° N. lat., and 77° to 80° W. long. It is bounded on the north and east by Lahoul and Spiti; on the north-west by Bara Bhagal, on the west by the native states of Mandi and Suket, and on the south by the Sutlej, one of the rivers of the Punjab.

The river Beas, another of the five rivers of the Punjab, rises at the top of the Rohtung Pass, 13,200 feet, and runs the entire length of Kullu. A mile below the source but a small torrent, across which one can jump, it becomes at Sultanpore, a turbulent river, used, when in flood, for conveying the

Government timber from the forests of Kullu to the plains below, where it is sold for sleepers. Unfortunately the water of the Beas, in its passage through Kullu, is so cold that there is not a decent fish to be found in it. It is said to be good for drinking purposes, and is so used by the natives, but as many of the dead of the country are thrown, half-burned into the stream, we did not affect it!

Three routes lead into Kullu. One is from Simla over the Jalaori Pass, 10,000 feet; one from Pathankote, and thence by the Kangra Valley, of tea fame, and over the Bhubu Pass, 10,000 feet; and another from Jullundhur, through the native state of Mandi and over the Dulchi Pass, 6000 feet, to Sultanpore, the capital. We choose the second mentioned. The distance, about 150 miles, is divided into eleven marches of very unequal length. We rode, and our baggage was carried by mules and coolies.

We put up at the government *dāk* bungalows, or rest-houses, which are provided at every stage. Anything more discreditable than their condition, however, cannot well be imagined. Furniture and crockery are conspicuous by their absence, and they are all in a bad state of repair, that at Sultanpore, the capital, being the worst; where it should be the best.* Yet a fee of one rupee for every twenty-four hours' stay is levied on each traveller.

The first part of the route is not very interesting, though the tea-estates through the Kangra valley are worth seeing. Not until we reached Dehlu did we get any sport. But there we made a bag of *chikor* partridge. At Jatingri, 7500 feet, we found plenty of pheasants, and there are some *goral*, and generally bears. From this onward pheasants abounded, but the *kalij* pheasant only. When we reached the top of the Bhubu Pass, however, we found *koklas*, *cheer*, and *manāl* pheasants. The great attraction of the route we had chosen was the beautiful march down the Bhubu to Karaon, through a superb forest.

* I must say, however, that a great improvement has taken place in 1892.

Kullu is famous for its forests. The largest deodar in India is to be found in the Solung nala. Oak, thorny oak, alder; three kinds of fir (called, locally, *tosh*, *rai*, and *kyle*); chestnut, wild pear, mulberry, elm, poplar, fig, birch, and rhododendron (which here grows to a big tree), abound everywhere. Besides these, there are five trees called royal trees, which are under the protection of the Forest Department. These, the deodar, the walnut, the box, the ash, and the elm, may not be cut down or even lopped without leave. Could but the difficulties of transport be overcome, what an enormous income the Kullu forest might yield! Government charges on the spot but three rupees for a magnificent deodar. But, unfortunately, the river Beas does not flood so heavily as the Chenab, the Sutlej, and Ravee. Consequently the difficulty of floating timber down to the plains is far greater than from forests bordering those rivers. Could the deodar, sold in Kullu for some three rupees, but be felled and cut up into railway sleepers on the spot, and floated down the stream, the sleepers alone would fetch from three to three and a half rupees each.

In Chumba there is a great industry in railway sleepers, and some years ago we paid an interesting visit to the Chatri forest, on a tributary of the Ravee. The trees are felled and cut up into sleepers, and then dragged to the edge of the high cliffs overhanging the stream. Formerly they used to be thrown over into it, but the process resulted in an enormous loss of timber. To overcome this difficulty, the Forest Department, a most valuable service, have made a sleeper shoot, some half a mile in length, zigzagging down the face of the cliff, and along which the sleepers travel by their own weight. At the end of each zigzag they are brought to a standstill by a mound of soft earth. At each corner a man is stationed, who instantly releases them, and starts them off down the next zigzag. The latter part of the shoot is made nearly level, so that the speed is decreased, and the sleepers fall quietly into the river, where they are collected and floated down till they meet the railway, and are there sold. Each sleeper is marked with a Government mark, and the penalty for stealing one is very severe.

Difficulty of transport also leads to an enormous waste of walnuts in Kullu. The natives, indeed, gather large quantities and grind them up for oil. But, as is the case in Kashmir with thousands of tons of delicious wild plums, apricots, apples, and pears, so in Kullu with the walnuts. There is no means of transporting them to India, where there would be a ready sale for them. The walnut oil is used for cooking purposes, and some of the European residents in Kullu assured us it was delicious, imparting a peculiar flavour to be got by no other means. Experience, however, of the odd dishes which old Indians have educated themselves into liking, made us somewhat shy of trying it.

We noticed on some of the walnut trees which were beginning to decay, an extraordinary natural ornament, which, later on, we discovered actually used to adorn European rooms in Kullu. The trees were covered with enormous fungi of a creamy French grey colour, flat at the top and with a wavy exterior, about fourteen inches wide by eight deep. They grow upside down on the trees, whence they are easily removed, and if placed against a wall form handsome natural brackets.

The Kullu natives, as indeed they ought to be, are well versed in tree-lore, and can name without any hesitation almost any species. But their superstitions are extraordinary. We noticed that they never passed under a tree which had fallen across the road without first placing a piece of wood or a stone on the trunk above them. I inquired into the meaning of this performance, and elicited the following explanation. There is a devil at the top of every tree. When it falls, he falls with it, and is therefore in unpleasant proximity to those who pass beneath it. In order to conciliate this wood-demon, an offering is placed on the tree before walking under it. After all, are there no people in London who decline to walk under a ladder?

CHAPTER II

WE reached Kullu on November 28th. As it was then too late for big game shooting, we rented a small house in the centre of the valley and went in for small game. The country is but the spurs and offshoots of the great range of perpetual snow which surrounds it on three sides. There are no low hills, and it is in these spurs above the narrow gorges and valleys which intersect them that the villages stand. It is a peculiarity of the Himalayas, the youngest of mountain ranges, that they rise so sheer and so abruptly, the valleys being mere ravines. All the villages are built after the same pattern, and laid out on the same plan, and the houses stand as close together as the nature of the ground will admit. Men and cattle herd together, and the filth of the surroundings is indescribable. But for the pure, crisp mountain air, and the delicious iced water everywhere abounding, surely it would bring a pestilence upon the inhabitants.

The fields which surround the villages are terraced on the mountain sides, the terraces being five, ten, twenty feet above each other, according to the steepness of the slope; and the principal crops are barley, maize, pulse, and millet on the higher, and wheat, tobacco, poppy, and rice on the lower ground. There are two crops in the year, one in the spring and the other in the autumn. The houses have three or four stories of one room each, and, as each story, except the lower, is surrounded by a verandah, they present a rather top-heavy appearance. The roofs, of slate slabs, or of split wood, slope slightly, and project beyond the verandahs considerably, giving the house a chalet-like look. The cattle live in the lower story, the family in the higher, and the verandahs are piled with crops of hay, rice, and straw, and of leaves on which the cattle live in the winter when the grass runs short.

Though picturesque enough at a distance, dotted here and there on the mountain slope, surrounded by green fields, or embosomed in forests of deodar and pine, and hemmed in by lofty snow-clad mountains, the villages do not bear too close an inspection. The people are as filthy in their habits as in their domestic arrangements, and seldom, if ever, wash. The men's costume in winter is a coat of home spun wool, called *puttoo*, reaching to within three inches of the knee; continuations of the same material tight below the knee; a round flat cap of black wool turned up all around, in the shape of a pork-pie, and grass-shoes, covering the front part of the foot only. In summer they discard everything except the hat and the coat, and when they wish, for courting purposes, to look particularly fascinating, add to the former a bunch of the lovely wild flowers with which the country abounds.

The Kullu ladies wear but one garment, and have no "fashions." It consists of a gray *puttoo* blanket. One end of this is brought over the left shoulder, passed behind the back, under the right arm, across the bosom, under the left arm, over the back again, with the end brought over the left shoulder. Two brass pins, connected with a coloured string, secure these ends, the pins passing through the strings and joining them to the main portion of the blanket in front. Their arms and legs are bare, and they wear absolutely no other article of clothing. For *melas* or other festive occasions, they make no addition to their dress—or should I say, undress?—but that of extensive gold and silver ornaments, only the rich however, affecting gold. This jewellery consists of bangles, armlets, ear and nose rings, finger rings, necklets, and bands of silver across the head. They prefer silver enamel to plain silver jewellery, and certainly their taste is to be commended, for it suits their colouring admirably. This enamel is made largely in the neighbouring district of Kangra, and though the work is coarse, the effect at a distance is very good. The women's poor ears are pierced all round the edges, and are simply loaded with rings, becoming an absolute disfigurement by being forced down by the weight of their ornaments. But what will women not endure for the sake of custom?—fashion I cannot call it, for it is unchangeable! Their head-dress is their prettiest adornment. It consists of a band of silver filigree work, with a fringe, and in front over the forehead, hangs a crescent-shaped ornament of enamel work also edged with a fringe. One of their many finger rings is almost invariably fitted with a piece of looking-glass the size of a shilling piece, and they pay particular attention to it, especially if they are at all endowed with good looks, as many of them undoubtedly are.

As in many foreign countries, the women are treated little better than beasts of burden. It is they who carry the loads, they who work in the fields, while their lords and masters sit about in groups chatting and smoking their *hookahs*. Along the roads the men ride the ponies, while the women trudge along with a baby slung across their shoulders and a huge load upon their heads. There is practically no marriage ceremony. The men barter with the parents of the girl they wish to make their wife, and thirty to forty rupees is the average value of a woman as a worker in the fields. These marriages, if you can call them by such a name, rarely turn out happily. The people are naturally immoral; it is not unusual for a woman to leave her husband for days or weeks at a time, and when she returns, or is brought back, she remains unpunished. The husband is only too glad to regain her services, and fears if he were to punish her she would run away again. The number of wives a Kullu man takes to himself depends practically on the amount of land he has under cultivation. The Kullu servants we engaged were frequently requesting leave to go in search of an errant wife, and many visits were made to them by others on a similar errand. The remedy appears to rest with themselves, but they do not mind; it is the custom of the country.

Housekeeping in Kullu was not an easy matter, as I very soon discovered. The people are a truculent race, who will neither work for you, carry loads, nor bring in supplies without pressure. The latter were by no means easy to procure. By the folly of the deputy-commissioner of the district of which Kullu forms a part, the tariff for charcoal, wood, grass, milk, etc., has been fixed higher than at Simla, itself the hill capital of India; and this in a country where the above mentioned commodities simply abound. Eggs and fowls, however, were especially hard to get, for the natives are a mongrel race of Hindoos, who think it against their caste to keep hens. In the way of drinking, however, we found no difficulty. Nature had provided an unlimited supply of the brightest, coolest, best water we had ever tasted, mixed with whisky or not.

CHAPTER III

AS may well be imagined in a country of altitudes varying from two thousand to twenty-two thousand feet above the sea, and uninhabited above nine thousand feet, there is plenty of room for sport. The game to be found in Kullu is as follows: Panther, bear (both red and black), ibex, serow, burrhel, barking deer, goral, ounce or snow-leopard, oorial and musk-deer; of pheasants, the cheer, kalij, koklas, argus, *manāl*; of partridge, the blackwood and chikor; also, duck, snipe, and woodcock.

Our bag for one year was as follows: pheasants, 137; *chikor*, 321; cock, 49; snipe, 9; duck, 3; barking deer, 7; *goral*, 3; black bears, 6; red bears, 8; and musk-deer, 3; and this might have been considerably increased, had we cared to go in for slaughter. With regard to some of the head, it is not more than would be shot in good coverts at home in one day. But consider the joy of shooting in these lovely mountains, in a perfect climate, where, for sporting purposes, the whole place belongs to you; where you take out no licence, pay no keepers; where the birds are bonafide wild ones, and take a lot of shooting; where you generally carry a rifle in addition to a gun, and run the chance of knocking over a bear or a panther, as well as a pheasant. All this can only be really understood and appreciated by those who have tried it.

Much is heard of the scarcity of small game in Kullu at the present time, compared to what it was a few years back, and often the European residents said to us, after we had had a good day, half in jest and half in earnest: "Why, you'll leave us nothing to shoot!" A moment's consideration, however, will convince anyone, that legitimate shooting, at birds fairly on the wing, and during the season only, would tend rather to improve than to deteriorate the stock, especially in the case of *chikor*, the cock of which, when the coverts are too near together and the birds too thick, fight incessantly and *à outrance*.

But it is not difficult to account for the scarcity of game, when one takes into account the manifold enemies the poor birds have to contend with. In the first place, throughout the valley there are literally thousands of *zemindars* (small farmers) to whom are granted licences to carry a gun, nominally for the protection of their own crops from bears and birds, and their flocks from bears and panthers; but, actually, for the destruction of small game. Every evening hundreds of pheasants are potted in the trees, and scores of *chikor* are slaughtered as they sit huddled to gether on a rock in the cold mornings. Immense quantities are destroyed in nets, caught in traps, or shot, as they are feeding along a narrow line of corn carefully laid down in a likely spot, and watched by a concealed sportsman(?). Then, during the breeding season, numbers of eggs are taken either to be eaten or to be wantonly destroyed. The birds are captured, especially *chikor*, and sold in the different bazaars, where they are purchased for fighting purposes, to amuse fat and lazy natives, who delight in watching the poor creatures maim each other, though they would rather give up their dearest relative than fight themselves.

Besides this, I regret to say, that there are some European residents in the valley, who either provide the natives with powder and shot to go and shoot for them, or else buy the birds when shot. In addition to all this, there are the birds' natural enemies, which abound—the fox, jackal, stoat, weasel, hawk, and kite. With such a host to contend against, is it surprising that the game of Kullu is diminishing?

But, for all those evils, except the last named, which presents difficulties, there are remedies to be found. In the first place, licences should be granted with a more sparing hand. They are neither required, nor are they used for the ostensible purpose for which they are taken out. The price also of the licence, now only fourpence halfpenny, might be raised. With regard to the employment of natives by residents to shoot for them, the remedy is in the hands of the latter. Finally, practical measures might be taken to suppress poaching in a country more poached, I believe, than any other in India.

Kullu is one of the very few places in India having a reputation for woodcock shooting. But during our first year in Kullu, D., shooting over every likely place, and bagging every bird he saw, only shot forty-nine. This was considered a good bag. But in Albania he had shot many more in one day! In years gone by in Kullu, ten or twelve couple a day was considered a good bag. That would now be thought a fair bag for a whole season. The birds have been getting less and less, and the residents, taking it for granted that the woodcock were birds of passage, have accounted for it by the difference in the severity of the different winters. If they find few birds, they argue that they have not been driven down by the snow. D., however, is strong on a theory to account for the diminution of the cock. Every likely place is walked over almost daily, and every *sahib* tries to get woodcock, with the result that ninety percent of the birds which come down the valley are shot. Thus each year there are fewer left to breed from, and if some stringent measures are not taken, Kullu will soon know them no more.

Now, as an object for sport, a woodcock can hold his own against any game-bird; for among trees he is a most difficult bird to shoot, and as a table delicacy he is unrivalled. In very few places in India are they to be found at all; and it would be a thousand pities if, in any one of these places, he were to become extinct. In Kullu it would be an easy matter to save them. As they are not exactly the kind of bird to give many chances for a pot shot, and as powder and shot are too expensive articles to be wasted on the risk of shooting a bird on the wing, the cock enjoy immunity at the hands of the natives.

Kullu is too far distant from a railway to suffer from an influx of winter sportsmen, and, therefore, there remains only, practically, the very few European residents, including the assistant-commissioner, forest officer, etc., to account for the cock. Should these few agree among themselves for a close period, I am convinced that Kullu would, in time, regain its old prestige as a ground for cock, and that the residents would be amply rewarded.

As regards snipe, I may mention that there are none in Kullu, except the solitary snipe. A good bag of these is twenty during the season. Duck are occasionally to be got as they pass up and down the river, to and from India and the Central Asian lakes, where they breed. The river, as it passes through Kullu, is too rapid to permit of their resting, which accounts for their being so seldom shot; but a Mr. D.—one of the residents, has now constructed a couple of ponds close to the river, and between it and his house, and the ducks settle in large quantities, and he is able to keep himself and his friends supplied during the season. If other ponds were made, above and below, there would doubtless soon be good duck shooting in the country. The land in the immediate vicinity of the river is all waste land, so that it is merely a case of flooding. I commend the suggestion to those who live in Kullu.

CHAPTER IV

THOUGH there were plenty of pheasants without going more than a couple of miles from the house, we were greatly hampered by a lack of dogs. These are an absolute necessity for getting the pheasant out of the dense jungle.

Now, we had brought two dogs out from England, and two others had been lent us. The lamentable tale of their tragic ends would curdle the blood of "doggie" people at home. Before Christmas Day, ere we had been even a month in the country, two had been killed by panthers. Before we had been two months in the country, a third met with the same horrible fate, and the fourth, a spaniel, "Nell," had a very narrow escape.

Kullu abounds in these spotted beasts—I am almost afraid to give them a name, for the grand old controversy among sportsmen as to the difference between a panther and a leopard will probably last till the gentlemen in question change their spots! Some hold that the difference is in the claws, those of the panther being retractile, like a cat's, and those of the leopards resembling a dog's, or vice versa. Others assert that Mr. Spots of the plains is the panther, and him of the hills the leopard. D. says there's no difference at all!

Anyhow, whatever they were, they goaded us to desperation with their depredations on our four-footed allies. Not only to sportsmen, but to the sheep-owners in a pastoral district like Kullu, are they an absolute pest. They offer no fair sport, as it is almost impossible to circumvent them. Whenever opportunity occurs they attack dogs, and the poor farmers suffer heavy losses all the year round by their constant depredations on their flocks and herds. They are insatiable, and if they get a chance will kill a dozen sheep at a time.

We knew of one panther which got into an outhouse where one resident kept his dogs, and destroyed the whole pack of five in one night! Bold beyond description, their valour somehow is only apparent when one happens to meet them unarmed. Finally we got into the way, when carrying only guns, of always having with us a couple of cartridges loaded with ball, but they never gave us a chance.

Poor Nell's adventure was as follows:—We were returning from shooting, and had sent on our rifles. D. was standing close to the edge of the jungle lighting a cigar, when he heard a cry. He thought it must be a monkey, as there

Panther Cub

were some about. But a sudden fear seized me that it was "Nell," and rushing up on to a boulder near, I looked about. No sooner had I reached the top, than my worst fears were realized, and I shouted down to Dick that a panther had got "Nell," and rushed off in the direction the beast had taken. From below D. could see nothing, but fired, knowing that that is the safest thing to do, as it frightens the animal and often makes him drop his prey. The shot had the desired effect. The panther dropped the dog, and sneaked off up the hill, where we could see him plainly; and, as there was still plenty of daylight, had we only had our rifles, Spots would certainly have run a good risk of making us a present of his skin.

Meanwhile, poor Nell rolled down hill, and brought up in a thorn-bush, after a clear drop of twenty feet. When we reached her she was more dead than alive, probably from fright; but she had a nasty gash in her throat, the place where the panther always seizes his prey. We got her home, and eventually cured her, which was lucky, as the bite of one of these beasts is poisonous to a degree, owing to the carrion on which they subsist when pressed by hunger. Their claws are equally venomous, and a scratch from them not unfrequently causes blood-poisoning.

Not many days after the spaniel's escape we were returning from shooting down the same ravine, but without the dogs, when we heard above us, on the opposite side, the short, sharp bark of a "barking deer." D. instantly dropped down and watched, and, ere long, the little animal came in sight, evidently in a terrible state of excitement, barking continuously, and stamping his feet. Dick might have had a long shot at him, but refrained, as he fancied the deer was being hunted, and probably by a panther, though occasionally two pine-martens, hunting together, kill these "small deer." He had not long to wait ere his friend came along, probably the same which had made a bid for poor "Nell." He sneaked through the brushwood, taking advantage of every bit of covert, like a trained skirmisher, and was within twenty yards of what he hoped would be his supper. D. waited till he came to a fairly open spot, and though he judged the distance so great that he could hardly hope to bag him, he could get no closer, and so fired, only, alas! to miss. Poor D. went home in anything but an angelic frame of mind, for when was he likely to get such a chance again!

One moonlight evening, not long after, we were returning from shooting about three miles from this ravine. I was riding on a yard or two ahead of D., when a panther crossed the road not five yards in front of me. He sprang on to a stone on the other side of the road, within five yards of us, and began bobbing up and down like a Jack-in-the-box, staring at us. But we were helpless. We had left our guns and rifles with the gun-carriers, and were hurrying home to dinner. This was probably the same panther, who had now escaped us for the third time. He had killed a wood-cutter before this, but he did not seem to take a fancy to us.

Panthers not infrequently attack men. A curious case occurred not long ago to an acquaintance, a neighbouring forest-officer. He was walking with a friend along a forest-road in the middle of the day, carrying an umbrella up to keep the sun off, when at a turn of the road they suddenly came upon a panther. The forest-officer, being unarmed, in order to scare him away, resorted to the device the lady applied to the tiger in the time-honoured picnic story, opening and shutting his umbrella at him. Spots, however, far from being alarmed, charged the forest-officer and, knocking him down, stood over him. The friend, meanwhile, had bolted (I suppose he had forgotten something), leaving him at the, mercy of the panther, who, fortunately, did not take advantage of his opportunity, but shortly after quitted the official, leaving him unharmed, but considerably upset in more ways than one. He described his feelings as the brute stood over him, how he kept wondering where the panther was going to begin upon him!

When one considers the amount of damage done by panthers, it is not surprising that every man's hand is against them, thereunto further encouraged by a Government reward for each one destroyed. Box traps are constructed, nine feet by eighteen inches, and with a bait at the further end, behind a partition, which is generally a puppy or young goat, or something warranted to give tongue. A sheepskin is hung on the bars, and when the panther, attracted by the noises of the bait, comes to make inquiries, and, reassured by the sight of the sheepskin, approaches it, he sets free the bar which holds up the door and is caught. The bait, quite safe behind the partition, is frightened, but not hurt; and the panther remains in durance vile till the trap is visited, and the muzzle of a matchlock, introduced through a hole in the top, gives him his "quietus." The skin is carried to the nearest Government official, in order to claim the reward. The latter cuts off the ears, to prevent the same skin serving a second time, and returns it to its owner, who sells it for the best price he can get, probably to a *fakhir*, by whom a panther skin is much prized.

We have often attempted to get a panther by sitting over a "kill." But it is disappointing work, for the panther rarely returns, and, if he does, it is generally so dark that it is impossible to draw a bead on him. A panther is a far more difficult bag to secure in this way than a tiger, and for this reason. A tiger, keen-sighted though he is, rarely looks upwards. A panther is more careful, and you have little chance of escaping his eye, however lofty your perch above him. Only once did we succeed in shooting one in this way, and now we never try it. The best chance of getting a skin is by tracking them, when the snow lies deep, to their den, whence they can be smoked out. Sometimes they are to be found lying in thick scrub, when they can be driven out and shot; but when they do bolt, their pace is so great—a succession of great bounds—that it requires a good shot to stop them.

A favourite form of circumventing panthers, much in vogue among the residents in Kullu, is by inducing them to commit suicide by shooting themselves with a spring gun! As soon as a panther has killed a sheep or a cow, and has been driven off for the time being, a wall of scrub, thorn bushes for choice, is built up round the kill. A small opening is left, facing which is a loaded gun, with a string attached to the trigger in such a manner that when the panther passes through the opening, he must inevitably discharge it and destroy himself.

They are the most difficult animals to get a shot at that I know. The only chance they give is when, themselves engaged in a stalk, their attention is for the time being distracted. One is then able to approach quite close to them. But how often in a lifetime does a sportsman get such a chance? Some men, however, are luckier than others, in sport as in other things. A friend of ours, when in Wardhwan, in Kashmir, had the good fortune to come across two white or snowleopards, hunting on their own account, and he bagged them both! These white leopards (*Felis uncia*) are most rare. The skins, grayish white with darker rings, the markings beautiful, are exceedingly difficult to procure. A good one is worth £20 or more, unmounted. Snow-leopards are generally found near the ibex ground, as it is on the ibex that they chiefly feed, and it is when stalking them that they offer the best chance of a shot.

CHAPTER V

DURING the winter months we solaced ourselves occasionally with a drive for game. *Hānk*-ing is good sport enough, if only the beaters will do their work properly. Naturally one only drives when there is a certainty of game, and it rests with the beaters to get it out. But, alas! you may take a horse to the water, etc. You can send Kullu beaters into the jungle, but you can not induce them to keep the line essential to a successful *hānk*. In Kashmir, where the people are Mussulmans, we have had some grand *hānks*; but here, in Kullu, the inhabitants are a degenerate race of Hindoos, and the biggest cowards in Asia. It is useless to expect them to keep line if a bear is anywhere within a mile of them. They move about the jungle in gangs, each trying to get the other to go in front, and utterly regard less of direction, while they decline to face the thickest and most likely covert.

Now there is absolutely no danger, unless from a wounded animal. All wild beasts will sneak away from a human being; and even a wild boar, probably the most dangerous of all, prefers to break back rather than face the spears, and does not notice any obstacle on his backward journey. Saving with pigs, in all *hānks* D. and I have ever made, and they amount to many score, we never but once knew a coolie touched, and that was when a wounded bear broke back and struck out with his paw in passing, removing a bit of the man's scalp. It is in following up wounded animals that numerous and fatal accidents occur, when the animal takes to some thick covert and makes a fight for it when approached.

Our first *hānk* in Kullu was up the Kois *nala* which had a reputation for holding black bears, besides *goral* and *khakur*, and our party consisted of five guns, including two ladies. The jungle was such a large one, that we could only take in half of it. The day before two hundred and fifty coolies were collected and sent to pass the night above the jungle. It was wet, and cold, and stormy, and the numerous fires that they lighted, to keep themselves warm, had the effect of driving away any bears the jungle might have held.

We made an early start at six a.m., and by 8.30 a.m. we were all in our places, and the *hānk* commenced. But, as usual, with so many beaters, they were absolutely under no control. Those who owned guns had brought them, a thing which should never be allowed in a *hānk*. Others had armed themselves with tom toms, which always do more harm than good. Others, again, perched themselves safely in trees, where they remained the whole time, and some appeared close behind us, very brave indeed, shouting and telling the rest what to do. Bang, bang, went the beaters' guns; lead flew about freely; the coolies were all over the jungle, and had a bear been afoot, it would have been impossible to fire. Disgusted, D. took the cartridges out of his rifle, and smoked hard, endeavouring to keep his temper. Someone, lower down the *nala*, got half a dozen shots at a *goral*, but missed, and the *hānk* was over. Then we moved on to another *nala*, Belindi by name, and the same farce was repeated—the bag for the day being one argus pheasant!

Our next *hānk* was at a place called Kelat, and again the same want of organization was apparent. We had sixty or eighty coolies, when a dozen would have been ample. The ground was a bare and precipitous hillside, intersected by small *nalas*, and studded with huge rocks on which were clusters of firs, under which the *goral* are usually found in daytime. The beaters were sent up two ridges of the hill about two miles apart, one half each way, and the guns posted in a *nala* midway between the two and near the top. It took us an hour and three quarters to reach our places from the river. Understanding that the *hānk* was to begin from the right, we faced that way and waited patiently for more than an hour. Then we suddenly heard a shouting behind us, and, on looking round, found the beaters from the left, and to whom we had had our backs turned, had finished their beat, and were now scattered about the hillside in our rear. Much annoyed, D. put down his rifle, and went to have a look towards the right, where, just over the ridge in front of us, he found a group of ten or a dozen men sitting smoking!

They had walked quietly through the middle of the jungle they ought to have *hānked*, and were now enjoying themselves till the fun was over, when they would come forward and demand their pay! D. pursued them, exasperated, and would have half killed any he had caught. But they were too fleet for him over bad ground, and, as all black men always appear to be exactly alike, he could not identify them afterwards; and returned to his rifle,

Dilaram, Kulu in Mid-Winter

and was about to descend the hill, when a fine *goral* came galloping down towards him from above. When he saw D. he turned off to the left, passing him at about eighty yards off. He fired, but, mistaking the rate at which he was going, missed. He disappeared downhill towards where I stood, and I killed him with a single shot. It was not a bad shot, as I was a long way off, and D. was not a little pleased, as it was my first "*goral*" As usual, the *hānk* had turned out a farce; but, by good luck, we had secured a fine head.

Once after this we were tempted to try a *hānk* for *kurt* up the Hamta *nala*, towards the Hamta Pass (fifteen thousand feet), leading into Spiti. There was the same want of control over the coolies, and the result was a blank, although D. had a long shot at a small *kurt*, which he missed. There were plenty of *kurt* about, and some grand heads amongst them; but, without system, it was impossible to drive, and would have been far more satisfactory to stalk them.

On one other occasion we did *hānk*, though as we had only eight men with us, it could hardly properly be called so. We knew where a bear was in a small *nala*, and collected the men. But as we knew, too, that the only chance of getting him out was to send them through the thickest part of the jungle very slowly, making no noise, but only striking the trees as they passed, we told them we wanted them to beat out a "barking" deer. D. remained close to them on one side of the *nala*, to see that his orders were carried out, and when they were about halfway through it out jumped the bear, making straight for where I was sitting, a little above the beaters. It was, however, a very hard shot for me, and I missed. Back came the bear through the beaters, and was out of sight before you could say "Knife," or D. was able to get a shot. But it was as good as a play to see the coolies bolt. No trees were high enough for them to swarm, and they had to be considerably reassured before they would descend again. We feared we should never be able to induce them to go out with us again.

On the last day before leaving our winter quarters, we let our rifle coolies walk through the Krarsu nala, and walked up with them, one on either side, and a little in front of them. We were fortunate, and succeeded in getting out three barking deer, two of which fell to D.'s rifle and one to mine. The flesh of the barking deer is not to be despised, so we went into camp with our larder well supplied with venison.

PART-TWO

IN CAMP ON THE SNOW-LINE

CHAPTER VI

THE winter of 1890-91 was as exceptionally severe in the Himalayas as in Europe. Snow fell heavily, even in March, and consequently bear-shooting did not commence till a month later than usual. On the 1st of April (auspicious day!), we went into camp about three thousand feet above our winter quarters. In Kullu, as elsewhere, there are favourite places for *shikar*, and the sportsman has to decide on which to take, and to watch carefully that no other *sahib* gets there before him. For the rule holds good throughout the Himalayas, that whoever first reaches a *nala*, or favourite hunting-ground, can retain it, for shooting purposes, as long as he likes to remain there, and no one else would dream of interfering with his rights. At the same time, etiquette would forbid his shooting in another place to the prejudice of other sportsmen. Should he leave the *nala* he has selected, to shoot elsewhere, even for a single day, any other *sahib* would have a perfect right to set up his camp there and take it from the first comer.

Attempts are sometimes made by the extra smart, or men new to the country, to pitch a tent in each of two *nalas* with a view to securing both. But the mere setting up of a tent gives a man no right to a *nala*—he must be there himself. Nor is it allowable, when racing for a *nala*, to send a tent on ahead. It is the man who must be on the spot, not his property.

This racing for *nalas* often begins from the very cantonments in the plains of India. Men, when the big game season opens, travel all the way together to Kashmir, as far as Srinagar. They leave Srinagar together, and then begins the tug of war. Perhaps the utmost that either can accomplish is four marches a day. They pitch their tents at night side by side, but, in the morning, one will probably find that his companion has decamped early and given him the slip. He, in his turn, will probably march all day and half the night to overtake him and get on in front, and thus they will go on till the coveted *nala* is secured by one of them.

Our canvas home for the next few months consisted of one Kashmir sleeping-tent, with bathroom and verandah; one field-officer's Cabul tent, for dining and sitting in, with bathroom for keeping stores in; three shouldari tents, singlefly, for our servants and cook-house. The total cost of our camp equipment was some four hundred rupees. D. carried a D. B. .500 Express, by Lancaster, and a hammerless D. B. gun by Tolley. I was armed with a .400 D. B. Express by Holland, and a D. B. .410 gun by Green of Cheltenham.

For those who desire to try a shooting trip in these grand mountains, a few words as to the dress I affected myself, and which I consider most suitable for ladies, will be found useful, and even for those who have no thought of such an under taking, it may prove of interest to know the kind of dress that is absolutely necessary for the work that I went through.

To commence with, I wore a very short plain skirt of the strong *khāki* drill, such as soldiers wear in India, and a Norfolk jacket of the same material. The skirt was not too narrow, or it would have interfered with the jumping and climbing over rocks that is so often necessary. On the legs I wore stockings with the feet cut off; on the feet short worsted socks with *puttoo* (homespun) over-socks, and grass-shoes. Bound round the legs I wore the grey *putties* of India, which are strips of *puttoo* two and a three-quarter yards long, by four and a half inches wide.

The advantage of wearing *khāki* is, that the colour is so admirably suited for sport, and that it never tears. Although it sounds cold, and undoubtedly is cold, that difficulty can be over come by wearing plenty of underclothing. A further advantage is that *khāki* can be easily washed, which was a consideration, as I had often to wash part of my own clothing. My rifle coolie carried across his shoulder, in a leather bag, a grey *puttoo* jacket, small cap and pair of warm gloves, which I put on when sitting in the snow. I myself carried my own cartridges in a leather pouch round my waist; but the coolie was intrusted with my field-glasses, and also, unless in the vicinity of game, always carried my rifle, though I never allowed him to touch it when loaded. As head covering when the sun was hot, I preferred an old double Terai hat of grey felt to anything else, on account of its portability, and because my hair was unsuited to the wearing of a *puggaree*. For camp I had a couple of short *puttoo* skirts, and *chaplies* (leather sandals) for the feet, and a cardigan jacket, which I often found very useful.

On the War-Path

At night the cold in the tents was often intense, and very warm sleeping jackets and pyjamas were essential. Inspite of putting all the clothing we could find on to our tiny campbeds, it was with difficulty we could keep warm.

Grass-shoes are absolutely necessary for walking with any comfort in the Himalayas. Each district has its peculiar kind. The Kashmir sandal requires a divided sock, with a partition for the great toe, as it is fastened on to the foot with a thin grass string, which passes between the greater and lesser digits, through loops right and left in the sandal, across the instep, and is finally secured by passing through loops attached to the upper part of the heel. In my opinion the Kashmir shoe is the best of its kind, for the string passing between the toes causes no inconvenience, and the shoe gives an exceptionally firm grip to the ground.

In Lahoul and Spiti the article is somewhat similar, but is made with a point turning up over the foot, something like the *chaussure* of a courtier of the fourteenth century. These shoes are lined with goat-skin, which is suitable for snow-walking, but not for wet ground, and one has to be prepared for both.

In the Kullu valley the shoes are made only five or six inches long, and are not intended to cover the heel. As these did not suit our requirements, we were obliged to procure them from Seoraj, some eighty miles lower down the valley, and experienced much difficulty in obtaining them. We invariably sent a pattern, to which, however, the maker paid no sort of attention, and out of a couple of dozen pairs we could seldom find more than three or four to fit. They are made of plaited ricestraw, and are shaped like an ordinary shoe, with a sole about three-quarters of an inch thick of the same straw. As they had no fastenings, we always carried a packing needle and thick string, and sewed them on securely, back and front, before starting. When they fit, grass-shoes are by no means uncomfortable wear, but they only last from two to four days.

A native *shikari* in the true sense of the word is unknown in all Kullu. A *shikari* is the native name for a hunter, and is a man supposed to know the country thoroughly, to be able to find the animal you are searching for at any time of the year, and, consequently, to know what they are feeding on at different seasons. A *shikari* must be able also to discover with his eyes game at a distance, and by his knowledge of mountaineering and his experience to guide you within shot. Here his share of the work ceases till the game is bagged, when it is his duty to skin it, convey it to camp, peg out the skin, remove all fatty substance, dry it by rubbing it with wood ashes and alum, and weak carbolic acid, and generally superintend the rough curing of it, until opportunity occurs to send it to the furrier. A really good man, capable of rendering assistance to this extent, is invaluable and worth a high wage. But such are rarely met with, though a few are still to be found in Kashmir and other parts of the Himalayas. In Kullu, however, they simply do not exist; and never could have existed, for the inhabitants are as devoid of sporting instincts as a Bengali *bābu*. We were obliged therefore to take a couple of coolies from the nearest village, to carry our rifles and show the way, for, on new ground in these mountains, without a guide, we should have been helpless. But all the work which should have fallen to the *shikaries*, even to preparing and drying the skins, we had to do ourselves, and very unpleasant work it was. I did my share, and cordially I disliked it. Fortunately, D. had had plenty of experience in stalking, and had also learnt how to prepare skins, so we felt independent.

Directly we left civilization and began camp life, I discarded promptly my Champion and Wilton saddle and my Busvine habit, and took to riding cross-legged in my shooting-dress, after the manner of the Kullu ladies. There is no doubt that the cross seat, in my Thibetan saddle, greatly added to my comfort on the bad roads (?) and was far less fatiguing on the rough and tiny ponies of the country, and with the perpetual up and down hill. D. purchased this Thibetan saddle from a Lahouli, and it was quite a curiosity in its way. In shape it is like a Turkish saddle, but the seat is made of "papier-maché" and is inlaid with ivory. We had great difficulty in getting the owner to part with it, as they are only made in far-away Thibet. When new it must have been a really good one of its kind, but when it came into my hands it had seen much service, was mended in places, and some of the inlaid ivory had come out. The seat was covered with a small Yarkand rug, and the saddle altogether was most comfortable, far more so indeed than D.'s, on which I had so often ridden long distances. I consider the cross-legged seat far preferable for mountain or slow riding to the side-saddle. With the Thibetan saddle it is an impossibility to give a horse a sore back, for the angle of the tree is very acute, and, instead of a single *numdah*, a couple of blankets or a blanket with one or more folds, are required to fit it to the back. The only danger is that of a girth gall from the native girth of raw hide and goats' hair, which is far too narrow for its purpose. D. always superintended the saddling, and my little mare never suffered.

Riding at all, however, was by no means always practicable. People in England have no conception what these mountain tracks about the Himalayas really are, and how one rides on the native ponies along places where in Europe one would hesitate to walk. These mountain paths are very treacherous, too, especially after a heavy fall of snow or rain. Cut out of the steep hill side, at best they are but a foot and a half wide, and an avalanche frequently carries away every vestige of them.

Road-making in Kullu is all done by *begar* or forced labour, and is consequently very ill done, the work being generally carried out by boys and girls from eight to fourteen years of age, superintended by a government *chuprassie*, who, seated on a convenient rock with an umbrella over his head, smokes a *hookah*! The mending takes place annually in October, after the monsoon, and in April, after the winter rain and snow. The wash of the rain off the mountain sides cuts large ruts in them in every direction. Landslips have to be repaired by men and women, but the only other mending attempted is done by the small children, with earth which they dig up by a small native hoe on the roadside, and carry in baskets holding about fifteen pounds at a time, and throw into the ruts. The first shower naturally washes the earth out again, and the road is of course as bad as ever. But what matter? The begar has been performed and the villagers cannot be called upon to mend the road again for six months. Everywhere are stones in tons, which are suitable for the purpose. But the use of them would mean increased labour, and it does not appear to have dawned upon Government to insist on the work being properly done. So the natives do it as I have above described, and give a finishing touch to their work by sweeping it over with a broom made of twigs. Thus are roads mended in Kullu.

Off the main road every man is his own maker and mender. Should you find a path destroyed you must either mend it yourself, make a detour, or go back. We generally contrived to mend, as, even if we had ponies with us, they could, like all mountain ponies, travel as fearlessly and safely as a man, over what, to the uninitiated, would appear impossible places. So good are these ponies, that one often trusts to them when it is not safe to do so, and that, not from any fear of their making a false step, but because the extra weight of a person on their backs may cause the track to give way beneath them. On one occasion we came to a place over which I wanted to ride, but D. thought the road rather rotten from the late snow, he insisted upon my dismounting, and had the pony driven over. When it was half way across, the road did give way, and down it went a hundred feet or more into the roaring river below. Fortunately the slope was not steep enough for him to turn over. He kept his feet in the most marvellous manner, but he could not stop himself. He slid on till he fell into the river, which was deep and in flood. Down the stream he was carried, sometimes with his head above, sometimes below, water, till we made sure he must be drowned. After being swept some two hundred yards, however, he touched the opposite bank, scrambled out on to his feet at once, and began grazing as though nothing had happened! Had the road given way with me on his back, however, the extra top weight would have caused him to turn over and I should certainly have been killed.

Not long after, D., too, had a narrow escape. We were riding along a mountain path after heavy rain, when suddenly the path gave way, and over he and the pony went, a clean drop of some ten feet on to a ledge of grass and mud, whence was a sheer fall into the river. The ledge saved the pony. It fell on it, and rolled over on to D., who found himself lying with his legs under the pony unable to move. Much alarmed, I shouted to the coolie who was with us to catch hold of the pony's hind legs and turn it over. The idea was a brilliant one, but anyone who had ventured within a yard of those hind hoofs would quickly have been made mincemeat of. The pony struggled greatly, and at length raised itself, so that D. could get clear, which he did with all speed and found himself safe and sound. He then seized the bridle and pulled the pony up, hanging on like grim death to prevent it going over the edge, and succeeded, though he was nearly gone once. A little further on he found a slope by which they were able to reach the road again, and we continued our journey, but the pony, though unhurt, trembled like an aspen leaf, and D.'s leg was black and blue for days.

Our choice of camping-grounds was small. The only flat places we could find were those in which sheep had been folded during the preceding autumn, and which literally swarmed with ticks. Only those who have been preyed upon by these fiends in insect form can sympathize with our sufferings.* These flat places are called *thaches*. They are open spaces, and when free from snow, look like green meadows, and are dotted about here and there through the forest. They are the likeliest spots in which to find red bears, as they are covered with wild carrots and rank grass, which those animals much appreciate. In the *thaches*, too, it is that Bruin, turning over the stones, finds the large white maggots, which he greedily devours. When the snow has melted, and the tracks of bears are difficult to discern, it is always possible to discover if they are about, by visiting these spots and noticing if any of the stones have been turned over.

We found the sheep a great bother. As the sheep approach, the bears go further away. During the dry season, in the plains, Lahoul affords grand pasturage for sheep. During the cold weather there is excellent feeding in the native state of Mandi. Kullu lies between these two countries, and enormous flocks of sheep are constantly passing through it on their way to and from Lahoul, during the months of April and May, September and October. During the latter season each sheep carries a little load of wool, and once amongst a flock I noticed a black *yāk*. *Yāks*, as everyone knows, are the long-haired, bushy-tailed, wild cattle of Thibet, and are very seldom seen at so low an altitude.

In the spring the upward-bound flocks are driven for feeding into the small *nalas*, which lead out of the main valley of Kullu, where they remain, feeding close under the snow line till the ice is sufficiently melted on the top of the Rohtung Pass (13,200 feet), to allow of their passing into Lahoul. Each flock is

* I have since heard of a remedy against the ravages of these troublesome insects; it is to rub round the ankles and wrists with "King Cocoanut" oil every morning, and not one of them will ever pass the oil.

accompanied by half a dozen huge sheep-dogs, which bark for twenty three hours out of the twenty-four. It will, therefore, be seen how exceedingly difficult it is to get clear of the sheep, whose presence is as incompatible with sport here as in a Scotch deer forest.

The *guddis*, or shepherds, are a queer nomad race, almost as noisy, with their ear-piercing sheep calls, as their big dogs. They live all the year round in the open air with their sheep. Most of them, I know, have never slept under a roof in their lives. They spend the winter with their flocks in Mandi, and the summer, fourteen thousand to fifteen thousand feet above the sea, in the pasturages of Lahoul. They are an exceptionally hardy race, never taking the trouble to erect for themselves any shelter from the elements. After three or four days' incessant rain, we have come across them soaked to the skin, and yet as happy and contented as if they were in a comfortable hotel!

The *zemindars*, or farmers of Kullu, are very glad to get the *guddis*' flocks to pass the night in their fields for the sake of the manure. As an inducement, they give free drink (*loogri*) and food to the *guddis* during all the time their flocks remain on their land. Drinking nothing but pure water for nearly the whole year, they are soon overcome by the intoxicant, and are often to be seen lying about the roads, hopelessly drunk, in the pouring rain, soaked through and through, but quite impervious to wet and cold. They invariably have a big dog with them on these occasions, which lies beside them until they are sufficiently sober to move on, and these dogs are very ugly customers to pass when on guard.

The dress is a short, very full loose *guddis* frock of *puttoo*, which, when fastened around the waist, has the appearance of a kilt, with an enormous goat-hair rope tied round their waists. Their legs and feet are bare, but their heads are well protected by a peculiar cap, made of felt, and which turns down to protect the ears and neck. The shape is not likely to find favour in Europe as a head-gear, but the *guddi* caps make quaint tea-cosies, I found, when new. These men are great smokers, but we could never make out what they smoked, as it neither looks nor smells like tobacco. Each *guddi* carries a *dach*, in shape like an English bill-hook, but without the wooden handle. It is all iron, and the handle is hollow, and they use it for a pipe.

CHAPTER VII

WE very soon regretted that first of April start into camp. April fools we had certainly made of ourselves. The weather was bad beyond description—a succession of storms of snow, sleet, rain, and wind, and the season a very late one. The snow lay feet deep in every direction, and it was very evident that there would be weeks ere it was sufficiently melted to allow of any grazing for bears. The discomfort of life in tents without any stove was great indeed. We were constantly encamped in snow, but, luckily, had brought with us a sufficient number of grass mats (afterwards replaced by tarpaulins) to cover the floor of the tents. We had a camp-fire, indeed; but we had the greatest difficulty in keeping it alight, as the wood for feeding it was always soaking wet. Necessity is the mother of invention, however. We discovered that an excellent way of warming a tent without damaging it, was to dig a hole in the ground beneath it, to cover it with a large flagstone, and keep it filled with hot ashes from the camp-fire.

A great source of danger in all *shikar* expeditions during the early part of the year are the avalanches, for which one has always to be on the lookout directly the sun gets hot. It is necessary to hurry through the narrow parts of the valley when there is heavy snow above, and, if possible, to pass them at an early hour in the day. When shooting near the snow-line, which, after all, is the only place for sport, one hears and sees avalanches, large and small, tumbling with terrific roars in all directions, and carrying with them rocks and trees by the tens and scores, which they hurl into the river below. Lives are often lost, and numbers of animals are annually buried in them.

We were lucky; but we had one rather close shave. We had gone up the Jagatsukh *nala*, to see if we could espy any signs of bears. The *nala* was a fairly broad one, and, no danger being apparent, about twelve o'clock we sat down to lunch. When we had finished, we moved on about fifty yards and then sat down again to have a look round with our binoculars. We had not been seated five minutes when we heard a tremendous roaring sound above us. There was nothing to be seen, but D. guessed what was up, as we were sitting on the edge of an old avalanche. With a shout to me and the coolies to make a run for it, we made off as fast as we could go, at right angles to the spot where we were sitting. Fortunately, the snow had melted in the direction in which we ran, so we were able to make good progress. Had we been obliged to run, or rather to try to run, over soft snow, escape would have been impossible. As it was, we were only just in time. An enormous avalanche of thousands of tons of snow came sweeping round the corner, on the very spot where we had been sitting, carrying everything before it. It was finally brought up by the bank on the opposite side of the *nala*, where it piled itself up into a huge mound which completely covered the pine trees (some thirty or forty feet high) under which we had eaten our tiffin. In our hurried flight, D. left his alpenstock behind him, and, when he returned to look for it, it had been swept away.

There are two seasons for bear-shooting—April to June, and from the middle of October till they hibernate, early in December. There are also two times to go after them—from daybreak till eight or nine in the morning, and from about four p.m. till dusk. Occasionally they may be found feeding in the middle of the day, but it is unusual. Now, to turn out of a warm bed at 3.30 a.m., with the temperature outside well below freezing point, would, I think, only tempt those really keen on sport. But the object is to get to a coign of vantage, whence can be obtained the best view of the neighbouring country, just before dawn, so that when day breaks the field-glasses can carefully scan every *thach*, or green spot, in the snow, which seems a likely feeding-place for "B'rer bar." There are other advantages in being thus early afoot. The snow is still frozen, and is easier to get over; whereas, later in the day, when the sun has melted the upper crust, one sinks knee-deep at every step. Moreover, in the early morning there is no fear of avalanches.

In spite of the weather and the non-appearance of the bears, we determined not to go down, but sallied out daily, whenever the weather permitted, hoping against hope that they would come out of their winter retreats, compelled to wander about in search of food, and thus cross the field of our binoculars. But it was not till the 30th of April that we saw the first bear of the season. It had been reported to us on that day that a black bear had been seen for some days near some hot springs about five miles from our camp. So we went up there, and waited about till nearly dusk. The cold was something abominable, and we were just about to give it up, when we espied our friend walking up the *nala* in front of us. As the wind always blows down, these ravines at night, D. followed the bear, and when there was just sufficient light to get a sight, got a shot at and bagged him. He was a small bear, but he had a splendid skin.

Six days later we heard of another black bear and two cubs, on the opposite side of the river. As it was some distance off, we moved our camp on the 6th of May to a spot near by, and were actually on our way to our tents, when we were met by a coolie, who told us that the bears were then feeding up a small *nala*, about half a mile above the place where our camp had been pitched. We instantly went after them, and soon saw them on the opposite side of the ravine. After deliberating a few minutes how best to get at them, we crossed over, and stalked them. When last seen they were feeding behind a slight rise in the ground, for which we made. When we had nearly reached the place, and, as it seemed a good chance for me, D. stopped, and signed for me to pass him and get first shot. But there was, unfortunately, an awkward piece of rock to be surmounted, and I handed my rifle to my coolie to hold while I crossed it. Hardly had I done so, however, when we heard a tremendous grunt close above us. Glancing up, D. saw the old bear facing us within ten yards, and looking nasty; so, without any hesitation, he fired, and the bear fell down the cliff within three yards of where we stood.

Thinking her dead, we now turned our attention to the cubs, who bolted along the ledge of an almost perpendicular rock. D. knocked them both over, one after the other. In the mean time, I saw the old lady, who had recovered herself in a marvellous manner, coming up the hill at us, and I fired, sending her head over heels down the mountain, towards the river. The usually mighty torrent was still entirely covered with snow, excepting in one small place, and into this hole, of course, as ill-luck would have it, the old bear fell. Dick descended and looked at the place, and decided that we had lost her. We went back for the cubs, picked them up, and placed them in a safe spot, whence we could send back and fetch them later. Then we set off again towards camp, bemoaning our luck at having lost the biggest of the bears.

We were walking along the frozen snow which covered the ice-bound river, and had just reached another hole in it, when an exclamation from the coolie made us look in front. "Sahib!" he cried. "Dekho!" And sure enough, just as we came up to the hole, out floated the old bear, still alive and kicking, and looking exactly like a large black seal. I fired again, hitting her in the back of the neck, and killing her. Luckily she floated to the side where the water was shallow, and we were able to seize and secure her at once. The whole business was a great piece of luck, for she was carried by the current under the snow for at least three hundred yards. Had she been dead, she would never have got through, but having some life left in her, she succeeded in scrambling over all the obstacles she met with on her way, and in floating out at the actual moment when we were passing.

Shortly after this we shifted camp to the Solung nala at the head of the Kullu Valley, and began to work hard for a shot at a red bear. Though we were up almost daily between three and four, returning at nine a.m., and out again at three p.m., and not home till dark, we saw nothing till the morning of May 16th. Then, just as it was getting light, we came across a huge black bear. He was traversing the snow at the bottom of a ravine we had just reached, and into which we were looking. But he was out of shot, and for some time kept wandering about in the aimless way that bears do when one is watching them, as if he could not make up his mind which way to go. At last, however, an idea came into his head, and he made straight in our direction. This was most fortunate, as we could not have gone towards him without showing ourselves. It was a grand chance for me, and when he got to within one hundred and twenty to a hundred yards, I fired. With a yell which would have done credit to John Peel, he began turning round like a teetotum tilt he fell, but recovered himself slowly, and moved off. Thinking it best to make sure, D. did not wait for me to fire again, but gave him a shot to which he succumbed; nevertheless, when D. fired, Dick said he was undoubtedly my bear, for he was as good as dead. Thus we bagged an exceedingly fine bear, which when pegged out for curing, measured seven feet one inch.

For some days after this we came across no game. But we did not much mind, for at this season of the year the scenery and the climate at these altitudes was enchanting beyond description. The mountain-sides were clothed with rhododendron—white, mauve, and scarlet, standing out against a background of grey rock and snow. Overhead a sky of intense blue, and an atmosphere clear and invigorating as champagne. Life altogether was worth living, and we were well content, though getting less sport than we had anticipated.

CHAPTER VIII

ON the 14th of May, we went down the valley to attend a *mela* or native fair. Nearly every village in the country indulges periodically in one of these festivities, which are all very much alike. A few days before the *tamasha*, men, women, and children from the surrounding villages begin to congregate at the spot where it is to be held. There are no booths or tents pitched, for none are needed. These people are a hardy race, and are quite content with the one country-made blanket, without which they never stir, wearing it on the hottest day, and having no more to cover them on the coldest night.

On the day of the *mela*, the tomtoming began at an early hour, announcing the approach of the different *devtas* (gods) which were to take part in it. They had been removed from their various temples, and each placed in a roughly made litter supported on poles and covered with coloured rags of every description, and decorated with the wild flowers of the season. Each was thus borne along on the shoulders of four stalwart villagers, preceded by the priest of the temple ringing a bell made of brass, iron, or tin, according to the wealth of the shrine, and accompanied by a native band of tom-toms, pipes, and the most ghastly sounding cow-horns it is possible to imagine. As each idol rejoices in the possession of such a band, and they all arrive on the spot at as nearly as possible the same moment, and as there are never less than twenty-five deities attending the fair, the noise, as may well be imagined, is so awful and so incessant that, having once seen a *mela*, D. and I have registered a solemn vow that nothing shall ever induce us to go near one again.

But for this abominable din, the scene would be picturesque enough. The colours are varied, the men picturesque, and, among the women, many are really good-looking. The gold and silver jewellery of purely local shapes and designs with which they are covered, is interesting to study. Through the upper part of each ear are hung their earrings, usually as large as bangles. As they wear as many as eight in each ear, the latter hangs down in anything but a pleasing manner, and the custom, I should imagine, is a painful one. The married women wear the Kullu nose-ring and the spoon-ring of pure gold, set sometimes with stones—sapphires, turquoises, and coral. Upon their heads are beautifully made silver bands of filigree work and silver chains, with numerous amulets of silver enamel; pendants are wound round their necks. This silver enamel is largely worked in Kangra, but it is not to be compared with that which is made in Mooltan. There is no buying and selling done at these fairs.

The people merely come to dance, drink, and amuse themselves. The men do the dancing, but both sexes are pretty good at drinking. The liquor is *loogri*, a kind of beer made of rice water, powerful, and very detrimental to the health. It is generally made by the Lahoulis and looks like dirty milk. Never having had the courage to give it a trial, we cannot say what it tastes like, but its effects appear soporific rather than pugilistic.

On our return from the *mela*, at four p.m., Dick at once started off to look for a bear. I had to stay in the tents, and make beds, etc., as none of our servants, who had had permission to attend the *mela*, had returned, so my temper was not of the best. Crossing the river, he ascended a small *nala* opposite our camp, and about a thousand feet above it, and, at a convenient spot, set himself down to survey the country with his glasses. He saw nothing; but the native with him soon marked a red bear and a couple of cubs feeding upon the opposite side of the *nala*. The sight of these hill-folk is simply marvellous. Nothing escapes them, and I would back their eyesight against ours when aided by the most powerful glasses. The bear and the cubs were not more than half a mile off, as the crow flies, but D. had to descend one side of the *nala* and pass along it, till he found a snow-bridge on which to cross the torrent. He got over at a spot about two hundred feet below where he had last seen them feeding, but could not get a glimpse of them. The wind was fair, and only care was needed not to expose himself to view. Presently he perceived them, but they had moved out about a hundred yards, and were now some two hundred and fifty yards from where he was lying behind a rock, and they were feeding head towards him. Had he but waited patiently, he would probably have got a fair shot within a hundred yards, but he foolishly accepted a chance when the largest one was within a hundred and eighty yards of him (as he afterwards discovered), and facing him, instead of being broadside on, and, in consequence, he missed. He was lying in a very unfavourable position for a second shot, and, before he could draw a bead on her again, she was off! Thus, instead of returning to camp with a fine red bear and, at least, one cub, he came back exceedingly ashamed of himself, and to receive a most severe lecture from me, for having been so foolish as to throw away a chance! But, at least, it was satisfactory to know that the red, or snow-bears, as they are called in Kullu, were at last on the move, and we looked forward to a good time coming.

Out at daylight on May 17th. We saw nothing, but went out again at three p.m., and, just before dark, I spotted a black bear across the river. The wind being fair, and the roar of the torrent drowning all sound, we had no difficulty in getting near him, concealing ourselves behind huge boulders. As he walked along grazing, I fired, but I did not allow quite enough, and only broke his hind leg. He bolted down the bank of the river, and we followed on the near side, and very soon caught sight of him crawling up the side into the forest. He was very sick, but D. was obliged to fire, and bowled him over. He only just escaped rolling into the river by a rock which stopped his downward progress. The coolies crossed by a snow-bridge, about a quarter of a mile further up, and removed the carcase to a higher spot, to prevent its being carried away by a chance flood, and then, as it was now dark, left it till morning. It was only a moderate sized bear, but its skin was good.

Up till now we had been able to make use of the snow-bridges to cross and recross the river; but, now that the sun was growing powerful in the middle of the day, we feared the snow would soon melt, so we began to make bridges. This we did by cutting down a couple of trees on the river bank, in such a manner that they fell across the stream, forming a ready-made bridge. The natives are very clever at this job, and can fell trees so as to fall within a foot or two of where they require them. The first bridge we made proved, subsequently, quite unnecessary, as we were able to cross by the snow-bridge till we quitted the *nala* on July 1st.

From the 17th to the 26th we had another spell of abominable weather—gales of wind, snow, and rain, instead of the bright sunshine due to us by rights at this season. For eight mortal days we were practically confined to camp. Any attempt to shoot was useless. In fine weather the wind blows, with wonderful regularity, steadily up these *nalas* from ten a.m. to four p.m., and downward after that hour, making stalking comparatively easy. On the other hand, in weather such as we now experienced, the winds blow from every point of the compass for five minutes at a time, so to speak, and it was simply impossible to get within half a mile of a bear.

During these storms, our servants' tent was blown down, and the tent-pole broken. As bamboos did not grow in this part of the world, we feared we had lost the use of the tent for good. We sent down to a resident in the Kullu valley, who kindly sent us a tent, but it had no ridge-pole, only a rope to answer the purpose, and would not have stood the weather we were having. Luckily, however, I remembered having seen, in an old bungalow down below, a bamboo that had been used for carrying hens in a crate. We sent for this, and D., with the tools which he always carried about with him, was soon able to replace the broken pole, and our servants were able to return to their own tent, and we to our dining-tent, which we had lent them. They much preferred their own, as we insisted on great care and cleanliness when they occupied ours, and, of course, allowed no fires in it.

About this time, we experienced great difficulty in getting coolies and supplies. The people of Kullu are a rich and truculent race, who have been greatly spoilt by the mistaken kindness of different administrators. Natives require ruling with a strong hand, and look upon leniency as weakness, and take advantage of it speedily. This has been the case in Kullu, where a military ruler is now required to bring the people to a sense of discipline. A strong man, strictly just, but firm—a man whose orders would be carried out to the letter, and who would himself see that they were, instead of trusting to *bābus* and understrappers, would soon put matters right, and prove a benefit to the country.

In other parts of India, a *sahib* is looked up to and respected by the people, because he belongs to the great *Sirkar* (Government). They respect the *Sirkar*, and know that, if it is not freely given, that respect will be enforced. But it is not so in Kullu, where the people have too long had their own way. We were obliged constantly to appeal to the administrator of the Government for the time being, in order to get supplied with even the necessaries of life. On one or two occasions the people of the nearest village refused to supply even milk or flour for our servants. For several days, too, we had been unable even to hire a coolie to carry our rifles, and it was impossible for me to manage mine in the difficult walking up the mountain-sides, for we thought nothing of an ascent of two thousand feet to get a shot. The assistant-commissioner kindly sent for the head-men and fined them, and we had no further difficulty, at least, not in that particular place.

It must not be imagined that a shooting expedition in a country remote from civilization is all play and no work. I have endeavoured to describe the rosy side of the picture only, but difficulties did present themselves, and not infrequently the difficulty of obtaining food would be enough to deter many from undertaking the trip. A greater difficulty to my mind, however, was that of procuring servants to accompany or to remain with us in this cold and inhospitable country. We started well, with a retinue suited to our exalted position(!); but we were only supposed to be going as far as Kullu. When, however, we signified our intention of prolonging our travels into Lahoul, the slaves all began with one accord to make excuse, and it was only by offering a large increase to their wages that we succeeded in keeping three of them. Now, one of these was a fool, another an idiot, and the third a convict, so it will easily be imagined how much menial work fell on my shoulders. To give an example of what one day's work meant, I will describe one on which we shifted camp, as a specimen of what the daily routine was when marching, sometimes for days together.

Rising between five and six a.m., according to the time of year and the state of the weather, we had, first of all, to dismantle our tents of all they contained, and pack the things into boxes. Then our clothes had to be packed; next our bedding—each article having its own particular place, and all being folded in waterproof coverings, not so much for fear of wet, but from the danger of their contracting some of the vermin with which all the hill men abound. The bedding disposed of, the beds had to be unlaced, taken to pieces, and rolled up, and packed each into its own case. The tables and chairs had next to be taken to pieces and tied together in a load, the rugs and tarpaulins then rolled up, and placed outside. I then would go off to the other tent and dismantle it, leaving D. to strike the sleeping-tent and to collect the tent-pegs, a most important point, which required personal superintendence, or else all the pegs, especially the iron ones, would have been lost in a month. Meanwhile, I was packing the tiffin-basket which accompanied us, and, after that, every article of crockery, glass, and plate, in use. Then the dining-tent had to be struck and packed, the three useless servants being interested with nothing but their own tents. It never took us less than two and a half hours to get under weigh, after we were called. We were obliged to see everything start on ahead before we started ourselves. If we passed a village on the road, we had to watch every load through that village, or we never knew when it would reach camp.

The march itself occupied about four or five hours, but we could generally manage to ride some portion of it. On arriving at the new camping-ground, we had to wait till the tents came in. A difficulty constantly arose by reason of portions of them coming in earlier or later, and retarding the pitching, as it was impossible to put everything connected with each portion on one coolie. When the tents were being pitched, the whole of the morning's work had to be gone over again in reverse order, and then I had to give out stores, order dinner, and see the meat cut up—a job I loathed. Meanwhile D. was superintending the digging of trenches round the tents, and when all was completed, and we had had our baths and some lunch, it was time to go out and shoot. If we saw any game, a long and often tedious stalk followed; and, game or no game, we never returned to camp till dark. Then followed the cleaning of our rifles, a work we never intrusted to any one, and then ensued a great washing of feet, an operation constantly necessitated by the wearing of grass-shoes, and then we dined, and by ten o'clock were quite ready for bed. If we had come across game in the evening which we had failed to get near, it was a case of getting up and after it at 3.30 a.m. the next morning. Even on off days, I had plenty of work helping to clean saddles (though my Thibetan saddle did not require much attention), making beds, trimming and refilling lamps, mending clothes, and often washing them, writing letters, and generally seeing to the cleanliness of the camp. From the above description it will, I think, be allowed that time was fully occupied.

CHAPTER IX

QUEEN'S weather! We felt especially loyal perched up hundreds of miles from such a thing as a birthday parade or *feu de joie*, and our loyalty was rewarded. On the 24th of May came a blessed change in the weather, and D. decided to go out next morning. I did not go out, as I had to prepare for sending a coolie to Sultanpur for stores and letters. D. left the tent alone at 3.30 a.m., and started up the *nala* in the dark. Just at daybreak a couple of red bears crossed the snow in front of him, at a distance of not more than eighty yards. But the snow was in hummocks, and he saw them but for a moment. He climbed to the top of the nearest mound, expecting to catch sight of them, but they had vanished completely. D. was intensely vexed, thinking he had lost his chance, but as it was still so early, persevered in following their pugs till he lost their traces in a large sunny spot where the snow had all melted away. As he thought it extremely unlikely, if they had sniffed him, that he should ever come up with them again, he took no trouble to follow up the trail.

About 6 a.m. he emerged from the forest, and reached a place which had been cleared by an avalanche, and there, not more than a hundred and twenty yards from him, he saw a fine red bear and cub feeding. D. was in a good place for a shot, and fired at the mother first; she turned over, fell on to a projecting rock, and then bounded off it like a ball some three or four hundred feet on to the snow below. The cub bolted up the hill, but Dick knocked him over with his left, and he fell into a small bush about ten yards below, where he remained. The difficulty was now how to secure the two prizes. They were only to be reached across a most dangerous angle of loose shale at such an angle that it was hardly firm, and over a smooth piece of rock, slippery as glass, from water trickling over it. He decided to go for the old bear first, and so went back into the forest, and made his way down till he judged he was about on a level with her. Then he emerged on to the landslip again, to see how it looked from there.

Fortunately, the bear lay not far off, and as the snow afforded a good foothold, they were able to reach her. They covered her with branches, as a protection from the vultures and the sun, and then they turned their attention to the cub. To cross the avalanche was impossible, but it appeared practicable to reach him from above, and to give him a push, as he was so small, which should send him straight down to the river, where he would lodge in the snow by which it was covered. D. had descended so much that the cub now lay fifteen hundred feet above him, and as the coolie was fonder of climbing than he was, he sent him off to do the job, and remained below eating snow, for he was very thirsty. The coolie got above the landslip through the forest, crossed to the far side, whence, by means of a ledge of rock, he reached the bear and gave it a shove. Down it came, but caught on another ledge below. This was not so difficult of approach, but caused further delay. Eventually, however, they managed to leave both animals together, and got back to camp about nine a.m. Four hours later, the skins were brought in. The cub was a year and a half to two years old, and was splendid furred; the mother, though much larger, was not in such good condition.

At three in the afternoon, we both started off for a place we had named Look-out Hill, about a thousand feet above our camp, and which commanded a grand view of the favourite feeding grounds on the opposite side of the *nala*. After sitting there for about two hours, I suddenly turned to D., saying, "I see a most enormous bear!" D. took a look at him through his glasses, and pronounced him a real beauty. He was well up the mountain on the other side of the river. It was cruel for me, that in nearly every instance, though they were my eyes that discovered the game, D. had to take the shot on account of his superior capability of getting quickly over the ground. D. thought he might reach him before dark, so set off at once at a run. After an hour's hard climb, he was delighted to see him feeding in the same spot about four hundred feet above him. The noise made by the mountain torrent greatly assisted his approach, and eventually he got within eighty yards of him, and drew a bead on his shoulder. He rolled over, but had been hit a little too far forward, and only his shoulder was broken. He made off slowly, but a second bullet settled him. This was a magnificent bear, with a coat simply perfect. We had the head mounted, and gave it away. D. had now made a record—three red bears in one day, and was not a little pleased with himself.

We began to have trouble about our skins. Some that we had sent down the valley to be roughly cured by a local leather-worker were returned to us on May 27th in the most deplorable condition. They were utterly destroyed. Instead of being stretched and dried at once, as they should have been, they had evidently been thrown on one side to await the shoemaker's convenience. So the flies and the insects had got at them, causing the hair, to slip. To hide these ravages, the *mochi* (native shoemaker) had cut away large pieces from the eyes, lips, nose, ears, and feet, totally ruining the skins. Evidently there was nothing for it but to perform the curing of the skins ourselves, and we had now three to begin upon. It was an exceedingly hard, dirty, and troublesome job. To cure a skin sufficiently to ensure its keeping till it reaches a furrier, will take hours for three or four days according to the state of the atmosphere, the size of the skin, and the amount of fat on it; for every scrap of flesh and fat has to be pared off with a sharp knife, wood ashes then rubbed in, and burnt alum powdered over it. When the skin is fairly dry, a solution of white crystal, carbolic acid and water (one part of carbolic to forty of water) sprinkled over it is an excellent dressing. As the skin is pegged out on the ground, all the work has to be done on the hands and knees, so it can well be imagined that it proves no child's play. How I hated the job! But D. said I worked with a will, and was a great help.

We were busy over our skins till the 29th of May, when we shifted camp further up the *nala*. That evening we saw a red bear, but did not get a shot at him. We went out after him again next morning and saw him, but could not get near him. That evening we went after him again, but in vain; and also on the following morning and evening, but we saw him no more.

In the early morning of June 3rd, we came across the track of a musk-deer which we followed across the snow. At last, looking over a ridge, I spied him sitting on the snow. He saw us, too, but, curiously enough, made no attempt to move, and as it seemed a nice shot for me, D. turned round to sign to me to shoot. But he saw at once by my disappointed face that something was wrong. A dreadful thing had happened. I had forgotten my cartridges! Words fail to describe the anguish of that discovery. So D. fired at once, and the deer bounded over the ridge. I declared he had missed it. Perhaps that was the green-eyed monster, goaded by my disappointment. But the coolies backed me up, though D. indignantly disclaimed the soft impeachment. Nothing, said he, would ever induce him to believe that he could possibly miss an animal at sixty yards, and sitting. He went to the spot to investigate for himself. There was no blood on the ridge where the deer had been sitting, but on peeping over, there he lay, stone dead, shot through the heart. Often as one has heard of animals performing wonders when shot through the heart, we our selves had never met with a similar case before, and were utterly at a loss to account for it. The musk deer is a valuable beast to shoot, as the musk-pod sells in India for from twenty to twenty-five rupees. The skin is useless, as by no possibility can the hair, or more properly speaking, the bristles, be prevented from slipping. The feet, however, make excellent dog-whip handles, and once Rowland Ward made me a very pretty pair of muffs out of them, removing the bone.

On the morning of the 3rd of June, we walked miles through the forest, thinking we might chance on a bear, feeding; but though we came across several fresh tracks, we saw no bears. After a hard morning's work we decided in the evening to go to our old place, Look-out Hill, and use our eyes instead of our legs. After about an hour I saw a bear opposite to us in the same *nala*, and at almost the same spot where D. had killed the large bear on the 26th of May. It was too far off, however, for me to hope to reach it before dark, as, of course, I could not travel as fast as D. Moreover, at the spot where the river had to be crossed there was no snow-bridge, and one had to make the best of a fallen tree which reached from one bank to the other. The water washed over it in places, and it swayed about in the torrent in the most uncomfortable manner. D. had tried it on a previous occasion and would not hear of my attempting it. The bank was so far perpendicular on the far side, that even he, with his long arms and legs, could scarcely clamber up. Occasionally, I feel the disadvantage of my five feet one inch of stature; but at other times I score, especially when creeping through jungles, where a tall man can only go with difficulty. My weight too, six stone six pounds, was often in my favour. It was hard luck, and D. commiserated with me, for really I worked as hard after the bears as he did, and took my share of the roughing and discomfort, though, often, when we came home at night, I was too tired to eat any dinner.

So D. started off alone, and after the usual climb, reached the spot where he ought to have seen the bear. He feared it had bolted, but continued up stream. He had not gone far when the rifle coolie touched him and pointed out the bear, feeding unconcernedly not thirty yards below. He fired, but he was too blown, in too great a hurry, and did not kill him outright. But the bear had to cross a large open spot in the snow, and received a bullet from his left before he reached the other side. All this performance I, from my vantage ground on Look-out Hill, distinctly saw. When I saw the bear move off after the first shot, I went through an agony of fear lest D. had missed him altogether. But the second shot reassured me. The bear was about three years old, and not a large one.

At last my chance came! Though out constantly, we had had no luck. On the 9th of June, after a hard and fruitless morning's work, seeing nothing, we were preparing, later than usual, for our afternoon start, about four p.m. A coolie came running up to our tent, telling us that some men D. had sent up the *nala* to fell a tree to enable us to cross the river at that spot, had seen a large red bear on the Dhundi *thach*, some five miles away. Now D. was sceptical about news brought in that sort of way, and thought it too late to go so far, on the chance of seeing nothing. So he decided to take another road and look for a bear, the tracks of which he had come across in the morning. He set out, imagining that I was going again to Look-out Hill. He did not see anything, and returned to camp after dark, when imagine his surprise and astonishment to find that I was not there!

Now, to say the truth, I had been exceedingly vexed that D. had not believed in the report of that bear, a report I fully credited, and no sooner had he departed on his own wild-goose chase, than I determined to go after it myself. So I summoned my rifle coolie, Doogloo, by name, and set off as fast as I could up the river bank. The ascent was not great, but it was difficult, over snow and slush, and much debris, to be got over or under, and which had been brought down to the river by snow falling from above. Moreover, there were two not inconsiderable streams to be forded. Over one of these Doogloo carried me on his back. But the greatest difficulty of all, was the necessity of crossing a snow-bridge, of which we had made use but a few days before, but which was on the point of falling in. In fact we had sent men up the *nala*, as aforementioned, in order to fell a tree to replace it. Poor D. remembered this shaky snow-bridge, when he found I had not returned to camp by dusk, and was very nervous about me, as he knew I must cross by it. It held, but next day it was gone!

Well, about six o'clock I reached the *thach* where the bear had been, but there was not a sign of him. I walked round it, and was about to give it up and go home, when I stopped to have one last look round. I then noticed something moving in the jungle on the far side, and called Doogloo's attention to it. He looked, but said it was a pheasant. This I would not allow, and I moved on a few paces, and then perceived that it was the bear, on his back, rolling! Now came the tug of war in the shape of the stalk. Without D.'s experience at hand to guide me, I had to arrange this by myself. But I managed all right and succeeded in getting within thirty yards of the bear. He was grazing behind a small hillock, and as I ascended it I saw him just below. The stalk had not been easy, as the wind that evening was changeable. But I had succeeded; I got my shot and knocked the bear over, but did not kill him. This is hardly surprising, however, when one takes into account how extraordinarily tenacious of life they are, and that I only carried a .400 bore rifle. But I followed up the trail through dense jungle, and eventually gave my friend the *coup de grace* through the head when within five yards of him. Fortunately he was so sick from the effects of the first shot that he did not go for me, as, in that case, Doogloo would assuredly have promptly bolted.

My trophy was brought into camp next morning. He measured six feet eight inches when stretched out for drying, and across the forearm seven feet. This was quite the largest bear that was shot that season, and he had a beautiful coat of golden brown, tipped with silvergrey. I kept his head, and sent the skin to Cheltenham to be cured and set up. D., I must add, professed to be extremely annoyed with me for venturing to run so much risk quite alone, but I fancy he was not a little proud of my performance.

CHAPTER X

ON June 10th, after being busy with skins all day, we started out late and ascended Look-out Hill. Almost directly we began reconnoitring I saw a black bear, but an immense way off down the valley, and D. did not think it worth while going after him. Shortly afterwards he, through the telescope, saw a huge red bear; but he, too, was at least two thousand feet above the river on the opposite side of the *nala*, and it would have been impossible to reach him before dark. He was the darkest coloured bear we had ever seen, and must have been very old. Instead of sitting still any longer D. suggested that we should walk through the jungle where we had come across the tracks of a bear the morning before. We found them again, quite fresh, and followed them for about half a mile, when all on a sudden we came upon the bear, feeding in a small *nala*, leading down to the main valley. He was some three hundred yards away, and we waited till he was out of sight. D. then wanted me to go ahead, but, as I could not move through the jungle as quietly or as quickly as he could, I made him take the lead. He came on the bear suddenly again, over a small rise in the ground; he was not more than twenty yards off, but bolted, and D. fired and killed him stone dead, his Lancaster bullet having passed clean through him. He was a small bear, but had a wonderful coat, almost white, and thus on this day we saw red bears with the lightest and the darkest coats we had ever come across. It was now quite dark, and we had two or three miles to go to camp, and a nasty steep bit through the jungle to reach the river bank.

Before we had proceeded far a considerable diversity of opinion arose as to which was the way. We divided, and D.'s section reached the river bank first, though further from camp than mine. They walked down the stream, and hearing no sound of us, began whistling with all their might. I may mention that for loud whistling these hill-men may be backed against the world. Half an hour later we came up with them, having had no small difficulty in the dark; but of their whistling and shouting we had heard not a sound, it having been completely drowned by the noise of the river. We knew our way now, but were obliged to proceed slowly and with caution, as we were not exactly on a turnpike road. About half a mile further we met the lantern that had been sent out from camp in charge of two of our menials, to meet and guide us. There were two of them, for I do not suppose that all the wealth of Solomon's mines would have tempted one man to venture in the dark alone! They are not afraid of wild animals only, but are great believers in ghosts, and terribly afraid of the dark.

We reached camp safely about 9.30 p.m., to find that there had been a disaster in the crockery department, and that nearly all our plates and dishes were smashed. We should by rights have embarked in enamel ware, but possessing a very common dinner service for which we knew we should get nothing when we returned to England, we thought we would chance it. The result was that in less than three months, our leg of mutton was served on a soup plate.

Next day we were so busy curing bear skins, writing letters, and sending clothes to the wash, that we did not go out shooting. This lastmentioned employment meant sending our linen a journey of fifty miles down the valley; but we looked forward somewhat to the day they departed, for when the coolie returned he brought us back some vegetables and fruit, great luxuries.

While we were sitting in the tent writing we heard a great "baa-ing" of sheep, and presently thousands passed our camp on their way up to Dhundi, the highest *thaches* in the *nala*. This appearance we hailed with anything but delight. For, as the sheep advance, the bears retreat. It is extraordinary that the shepherds of those immense flocks decline to sell even one sheep, unless compelled to do so, and then a misguided Government allows them to pick out the oldest and thinnest of their flock, and put their own price on it. Small wonder it is that the sheep are thin and tough, for they are never allowed to rest, but are kept on the move from morning till night. There may be a reason for this, though we could never discover it, for pasture is plentiful everywhere.

Early the next morning we sallied forth to see if we could meet again that dark-coloured bear we had seen so high up on the 10th, even beyond the birch forest, which grows nearer the line of perpetual snow than any other tree; but we returned without having seen him.

We now had a spell of bad weather again, and got no further chance till the 17th, when D., with a telescope, sighted a bear and two cubs, and, going after them, got a fair shot at the old one, within a hundred and fifty yards. But he was blown and shaky after a long pull uphill, and missed clean! This was now the second time. He lived to get yet another shot at her, and she lived to see us leave the *nala*. The coolies declared that it was perfectly useless to fire at her, because, according to their superstition, the days of all living creatures are numbered, and, if the number was not completed, it was impossible to kill them. In this case they really believed it, as they had seen D. kill some wonderfully long shots. But this bear he could not touch.

CHAPTER XI

ON Waterloo Day we shifted camp higher up the *nala* to Dhundi. The spaniel we had brought with us from England, had recently presented us with nine pups, all of which we kept till they were old enough to dispose of. When we marched they were carried by two coolies in our zinc bath, with a tent-pole passed through the handles, and covered over with a hammock to prevent their falling out. Poor little beasts! they made plenty of music en route, and did not seem to appreciate their ride. Ultimately, we gave away three, sold three, and kept three; the latter proving capital dogs, and retrieving well at seven months old.

We made an early start, and reached our new camping-ground in time for a late breakfast. Dhundi was one of the most beautiful, if not quite the most beautiful, of all our camps in these lovely Himalayas. Imagine a glorious green meadow of some twenty acres, carpeted with wild flowers of every colour, lying about eleven thousand feet above the sea, at the fork formed by the junction of the Solung river with the Beas Kund. From our tent-door, looking south-east, we had an exquisite view of the river after its junction, tumbling down the valley through a maze of grand chesnuts, walnuts, firs and deodars, to join the Beas proper, some thirty miles below. From our heights we could watch its course for miles, and, at that distance, enjoy the roar of its waters, as it fumed, and tossed, and struggled, onwards and downwards, to join the mighty Indus, when, wearied of its rollicking life from the parent glacier to the burning plain, it is carried onwards through the desert of Sinde, a sleepy, muddy stream, to its grave in the Arabian Sea.

Behind our canvas home rose giant snowpeaks, twenty thousand to twenty-two thousand feet high. Up the valley, to our right, we had a glorious view of the magnificent glacier whence the Beas Kund takes its source, a wildly grand and impressive spectacle. To the left, the valley of the Solung was still, even in the middle of June, completely bridged over with snow. At its head lay a pass often used, in less severe weather, by shepherds going to and from Lahoul. It was in this lovely spot, which no words can adequately describe, that we camped for ten days, enjoying the most perfect and cloudless weather, and the additional advantage of a grand full moon to illuminate scenes around us. Never shall we forget our camp at Dhundi.

The evening of our arrival there we went out up the *nala* running towards Lahoul, where we could walk with perfect safety on the snow which still covered the roaring torrent, and which must have been, in places, twenty to thirty feet deep. After walking about a couple of miles, we sat down to have a look round. It was piercingly cold, and it was here, I believe, that one of our coolies took the chill which necessitated his being sent down the valley a few days later, suffering from congestion of the lungs. I feared he would die before we could move him from camp, but he reached the hospital at Sultanpore safely, and we afterwards met him comparatively well.

We did not remain seated long, but started off to return another way, and very shortly came upon a bear feeding a few hundred feet above us. There being no covert anywhere about, it was necessary to approach him crawling on hands and knees, and, as this was a performance for which D.'s costume qualified him better than mine did me, to him again the stalk was intrusted. Never had he stalked a more suspicious bear. It was incessantly on the look out—a bite and a look round, another bite and a look out; but, each time it put its nose down, D. got a few yards nearer. When within a hundred and fifty yards, however, all nearer approach was out of the question, and D. lay down with his rifle ready for a chance. He waited and waited for this chance, till poor I, perishing with cold below, thought he was never going to fire. But the bear was either looking straight at him, a chance he would not accept, if he could help it, or, when broadside on, was just hidden by a small bush behind which it was feeding. From this spot it would not move, and, at last, D. was obliged to take the shot when the bear was facing him. It fell stone dead, and, upon examination, we found that the bullet had struck the left side of the head and passed out at the right side, which, as it was facing him when he fired, proved that it must have turned its head just at that moment, or he would have missed it.

On the way home D. shot a musk-deer, so we had done pretty well for the first day in a new camp. But the two stalks had taken a great deal of time, and it was quite dark before we had secured the musk-deer to carry with us, and we had a descent of fifteen hundred feet to make over a dangerous piece of country. After many falls and some hard knocks, we reached the tent at half-past nine, quite ready for dinner. But I was somewhat over-tired, and could not manage the incessant mutton as well as usual. No other meat was procurable, and it was a case of mutton, in some disguise or other, for three hundred and sixty-four days out of the three hundred and sixty-five.

On the 17th of June we were busy with skins all the morning, and in the afternoon went out shooting up towards the glaciers at the head of the Beas Kund. Hardly a mile from the camp we caught sight of a bear on the snow, about three hundred yards to the left front, and instantly dropped to avoid being seen by it. But for the intense cold we should much have enjoyed sitting on the snow watching the antics of the queerest bear I have ever seen. We called him "the mad bear." For more than an hour and a half we sat watching him, expecting him to leave the snow for a nice green patch which lay between us and him. But no; he kept roaming from side to side, first up and then down, apparently devouring great pieces of snow. He would lie down for a bit, then move slowly up the snow, only to squat on his hunkers and toboggan down again. Then he would gallop off about fifty yards in one direction, and then, after a moment's rest, gallop back again; in fact, playing in the most insane manner that I have ever seen a bear play in before and since.

But as it was growing dark and we were nearly perished with cold, and he showed no intention of coming down, D. set out to stalk him, though with scanty hopes of success, as he had to pass over snow the whole way to approach him. Fortune favoured him, however. Hardly had he begun his stalk when the bear moved off behind a snow bank, which completely concealed him. Thus, without much further trouble, D. was able to get a shot at him and bag the "mad bear's" skin. It was by no means equal to the others we had secured, and we perceived the time had arrived when the bears begin shedding their winter coats. So we decided to give up shooting them and to move slowly down the *nala* to Manali, at the head of the Kullu valley, where we meant to pass the monsoon. We had secured fourteen bears and three musk deer, one of the largest bags that had ever been made in Kullu.

My Trophies

From Dhundi we descended to another *thach* or meadow, called Jalari Age, where we remained a few days. On the evening of the 25th of June we went up the Pindaree *nala* and saw a couple of musk-deer, one of which D. got a shot at and bagged. This Pindaree *nala* is by no means a satisfactory one to shoot at this season of the year, for the entrance to it is very narrow and very steep, and is completely choked with snow and debris from the heights above. Day by day the snow melts gradually, and, in consequence, enormous boulders and trees are being continually released from its grip, and come tearing down the *nala* across the snow, to the great danger of any one ascending it. Never did we negotiate this Thermopylæ without many a warning shout of "Khabardar!" ("Have a care!") and some narrow escapes.

When we were up the *nala* after the musk-deer we again caught sight of our old friend the bear with the two cubs. They were feeding far away on the other side of the river, and as the shot D. fired at the musk-deer was too far off to disturb her, we determined, as, on our way home, we cast a last look at the old lady, to go after her on the morrow. We made an early start, as we had a stiff climb before us; but we did not reach the meadow where we had spotted our friend feeding till four p.m. But she was nowhere to be seen. We kept on to leeward on the edge of the forest and mounted up and up, hoping every moment that she would emerge for her evening's meal; for we always noticed that bears with young ones come out to feed much earlier than others. We fancy, too, that, when there is light enough, they graze all night also; therefore, when the nights are dark there is a better chance of seeing bears early in the evening than when there is a moon.

After a while I sat down to have a look round, and D. continued up the hill for another three or four hundred yards, when, over a slight rise in the ground, he came upon the objects of our search, feeding intently the far side of a partially uncovered *thach*. They were in a good spot for an easy shot, so D. dropped down the hill a little and signalled for me to come up. I hurried up as fast as I could, and we crawled on to behind a bush within a hundred yards of them. As D. was slightly in front he drew a bead on the old lady, who was broadside on to him, but as he expected me to fire every moment he would not shoot. But I could not think of depriving him of making sure of a bear which he had already missed twice. He did not understand this, and waited and waited for me to fire, till at last the bear moved away and was hidden from view. There was nothing for it now but to follow her up till she got behind a small belt of silver birches. When D. reached them he waited a few minutes and then went after her; but when he got through them she was nowhere to be seen. He looked round everywhere for her, and coming to the conclusion that she had smelt him and bolted, returned to me.

Now what really had occurred was this. As D. came out of the trees on the far side, the bear emerged on the near side, and continued feeding at a short distance from where we had originally seen her. As the wind was blowing straight down the *nala*, D. did not think it possible that she had gone back, but imagined she had scented him. Consequently, when he came out of the birches, instead of looking round as he should have done, he walked straight down to where I was sitting, and left the bears behind him! But, now, alas! they did see him, and bolted forthwith, passing about two hundred yards below us. D. noticed nothing, and would have known nothing about it had I not stood up and pointed. He instantly put up his rifle and fired; but it was too quick and difficult a shot, and he missed the old lady for the third and last time. Verily, she bore a charmed life, for when D. was looking for her she was feeding quietly within a hundred and fifty yards of where I sat, and I could have killed her at any moment, but imagining that he saw her too, was scrupulous about taking his shot.

It was now growing dark, and we tried to find our way home by a short cut. In consequence we lost it altogether, and had to wander about for hours. More than once we had almost made up our minds to sleep in the forest, but as it was very cold, we should probably have got fever, so thought it safest to keep on the move. Fortunately that day we had taken a shepherd with us instead of a coolie, and if anyone could find his way about these mountains it is a shepherd. Him we sent on in front, and followed slowly and carefully, reaching camp eventually at a very late hour, torn and dishevelled, but with no further mishap than the loss of D.'s Terai hat, which fell into the torrent and was washed away under the snow. In these latitudes the sun has great power, and an insufficiently protected head means sunstroke; so this loss was an important one, and next morning we had to send four hundred miles to replace it. Hitherto D. had generally worn a *pugree* turban when shooting, but the tradesmen down country had sent him up such gorgeous things in French grey and gold, that they would have frightened away all the bears in the country, so he fell back upon the common or garden Terai.

On June the 27th—day of days—after months of perpetual mutton, we received a present of a York ham! D. was desirous of eating it raw upon the spot, but I insisted that it should be boiled first. After some demur he agreed, on the condition that it should be cooked instanter. But I announced that it would need soaking for twenty-four hours. D. had to give in, but implored me to lose no time about it. Imagine his disappointment when I came to tell him that not one of our cooking pots was large enough to hold it! You might have knocked him down with a feather. He suggested our bath—already referred to as the travelling carriage of Nell's pups. But this I would not hear of. He then proposed digging a hole in the ground and putting a fire all round the ham, but this was deemed impracticable. So he gave it up, and finally had to send fifty miles to borrow a cooking-pot that would contain it!

That evening we went out again, and again we saw the old bear, feeding not half a mile from the place where D. had fired at her the day before. Plucky old lady! Long life to her! She was a mile away as the crow flies, and D. could not have reached her before dark. Had he imagined it likely that she would show again for three or four days after being fired at, we would have gone up the other side of the river to look for her. But she evidently had a poor opinion of D. as a shot!

On the 28th of June, we went up the Pindaree *nala* again, and after several escapes from the falling debris while passing through the gorge, D. succeeded in getting a long shot at a musk deer and killing it. On the 29th, we sent off twenty-four coolie loads of camp kit, and on the 30th, followed ourselves with twenty-four more, for Manali, four thousand feet lower down the *nala*, where we intended to stay during the rains, which were now due.

PART-THREE

MONSOON AND WINTER IN THE KULLU VALLEY

CHAPTER XII

WITH the end of June comes the end of the big game shooting throughout the Himalayas, and we had nothing to do but to remain quiet till the 15th of September, when the small game shooting would begin. We remained three weeks at Manali, and then went down still lower to Nugger, the rajah's old capital of Kullu, where the assistant-commissioner resides in the castle. He was about to go to Spiti on duty for a few weeks, and had kindly placed his house at our disposal during his absence. Nugger lies only five thousand feet above the sea, and after some months' sojourn in the snow regions, we found the climate there intolerably hot.

The castle stands on an eminence overlooking the Beas, and more than 1000 feet above it, and commands a grand prospect over the surrounding country, with a never-failing view of "Gaphan," a snow peak beyond the Rohtung Pass in Lahoul, twenty-three thousand feet high. Many legends are connected with this old castle, built nearly a thousand years ago, and, of course, it boasts a ghost! The story of the apparition runs as follows.

Centuries ago, one of the rajahs then occupying it gave an entertainment on the grassplot in front of the castle. This entertainment consisted of wrestling matches, performing goats, and other animals, then, as now, beloved of natives, and was viewed by the rajah, his ranee, and his courtiers from the verandah thirty feet above the lawn. While the games were going on, the rajah turned to his consort and asked her whom she considered the handsomest man there, intending, I suppose, that she should specify himself. But she pointed out one of the wrestlers, a fine, handsome man, and commenced to dilate on his charms. Forthwith the jealous rajah fell into a perfect paroxysm of rage, and ordered his Wazir that the wrestler should be beheaded on the spot. This was immediately done, before the ranee's very eyes. She, shocked and horrified with the rajah's cruelty, and distressed for the man she had so much admired, rushed round the verandah to the far side of the house, which over hangs the cliff by three or four hundred feet, and flung herself over the balcony into the abyss. She now, of course, "walks" at night, in the above-mentioned verandah, but did not honour us with a visit during our few weeks' stay at the castle.

This castle, itself old, replaces a still older one, which stood a few thousand feet higher, at "Chajogi," and was burnt down. The legend of its destruction runs thus. The reigning rajah had two wives, one of whom had presented him with a son. Not contented, however, he took unto himself two more, and appears to have been cordially detested by them all. The latest acquisition, indeed, took such a dislike to him that she determined to murder him. Consulting with the other wives, she laid the following plot. As it was desirable to continue the *raj*, and, as only one of them had a child, it was decided that that one should escape with her son from the castle, and that the others should perish with their lord and master, in the approved Hindoo fashion. Accordingly, they closed and barred the only two entrances to the building, and, during the night, set fire to it at both ends, and were all burnt alive. The castle was never rebuilt, but a new one erected at Nugger. The story goes that when the young rajah came to man's estate, and heard the story of his father's tragic death, and how it had been planned by his youngest wife, he issued an edict that all the women of the village whence she had been taken, should, for all future, be compelled to wear a large iron ring suspended from the left ear. Strange as it may appear, though over a thousand years have passed since this order was promulgated, it was almost universally acted up to till within the last few years, when pressure was put upon the Government to revoke it, and stop the practice. At the present day, however, I am told, it is not unusual to see many women of that village still wearing the badge of the disgrace of their kinswoman of centuries ago.

In the courtyard of the castle we noticed a small square structure, which we might have taken for a duck-house, had it not been pointed out to us as a temple. It contains a large slab of stone, some six feet square and four feet thick, which is said to have been deposited there by wild bees. The slab is an object of much veneration to the people about, who throng the courtyard in great numbers, placing offerings upon it of rice, meal, and wild flowers. The little temple was formerly the scene of a curious custom for settling legal disputes. Two goats, one the property of the plaintiff and one of the defendant, were placed in the temple, and the Brahmin in charge sprinkled water on each. The case was decided in favour of the owner of the goat who first shook the water off. They say, not so many years ago, this method of settling a dispute was resorted to by the assistant-commissioner for the time being, under the following circumstances. Two old women had a quarrel about some jewels, each accusing the other of having stolen them. They did not appeal to the commissioner in court, but persistently annoyed him to settle the case for them informally. To get rid of the importunate widows (?) the official bethought him of the time-honoured goat-test. He tried it; and whichever goat shook first—I forget which it was—the owner of the other said it was evidently the pleasure of the god of the temple that she should lose, and both departed satisfied!

CHAPTER XIII

EARLY in September, during a break in the monsoon, we made an expedition up the Rohtung Pass, which is the main route between the Punjab (by way of the Kangra valley) and Ladakh and Yarkand. It is practicable for ponies and mules, but the chief beasts of burden using it are sheep and goats, which carry saddle-bags, and are loaded with about twenty-five pounds.

The Rohtung Pass, thirteen thousand two hundred feet above the sea, is generally open from June to December. At the summit there is a level piece of path for about two miles, and, as the mountains rise on either hand to eighteen or twenty thousand feet, it forms a perfect funnel through which the wind rushes, at certain seasons, with such terrific force, between the hours of ten a.m. and four p.m., that it is often impossible to use the road at that time. Violent storms also accompany these gales, and make the pass a very dangerous one, on which lives are lost every year. On one occasion no less than two hundred and fifty men were buried under the snow and perished, and this too, I think, in the month of May.

We started from Manali, our then headquarters, for Palchan, the road lying along the left bank of the Beas, and running through fine forests of deodar and pine. Three miles on the road we came to the village of Bashist, celebrated for its hot springs. This, and two other villages in the valley, Manikarn, on the Parbatti river, and Kelat, on the Beas, are much resorted to as places of pilgrimage, and for bathing purposes by persons suffering from skin-diseases. At Bashist the springs are walled round, though open to the sky, but there are different partitions for men and women. Close by is a cold spring, and one can

Manikarn, with Parvati River frozen over

stand with one hand in icy cold water, and the other in water so hot that one can hardly bear it.

At Palchan the Beas is crossed by a wooden bridge, and the road continues on the left bank of the river. At this point the pass actually begins, and from the bridge a zigzag path ascends a precipitous height of eight hundred feet, to a cultivated plateau, now heavy with crops, and presenting a very different appearance to when we saw it six months before, covered with atleast ten feet of snow. Here we caught a glimpse of Rahla, some three miles distant, and fifteen hundred feet above us, and lost sight of the Beas, which now, for two miles, flowed through a deep narrow gorge. We crossed it again at the entrance to this gorge, and followed its left bank through beautiful forests till we reached Rahla, the last stage on the Kullu side of the pass. The hut here, placed at the disposal of travellers, and for which a fee of one rupee a night per head is levied, is a disgrace to civilization. But we were obliged to put up with it.

Rahla stands in the middle of the narrowest part of the ravine leading up to the pass. The fierce winds that blow so constantly on the summit are here severely felt, and even in the warmest weather it is horribly cold. There ought, consequently, to be a comfortable house for the use of travellers instead of a pigsty, in which it is impossible even to light a fire. We were rather excited at the idea of getting a panoramic view of a (to us) unknown land from the top of the pass on the morrow, though I do not think our excitement was so great as that of a man I know of, who took his wife home from India for the first time, and with her their only child, a baby. So excited were they that, when they reached Brindisi, they rushed on shore and took their places in the train before they remembered, or were reminded, that they had left their baby on board!

We quitted our comfortable (?) quarters at Rahla at six a.m. on the morning of the 4th of September, as we wished to have reached the summit and be well on our way home before the gale commenced. It was a dark morning, and we trembled for the views we had come to see. We calculated it would take us quite three hours to reach the top, and we found that we had timed ourselves to a moment. The road is not difficult, but it is a continuous zigzag throughout, which had this advantage, however, that there was no necessity to halt to look at the views which continually succeeded each other. There was no snow on the way up, but at the summit it was still in places many feet deep. The wild flowers were perfectly lovely, enjoying their brief span of life, between emerging from the snow in August to be buried in it again in October.

At nine o'clock we sat down to breakfast on the lee side of a rock. Before us spread the great waste of rock and ice known as Lahoul, uninhabited during nine months of the year, excepting at a few favoured spots on the banks of the Chandra-Bhaga. During the three summer months, however, from the middle of June to the middle of September, the mountain sides are covered with large flocks of sheep and goats, driven thither by the shepherds of Kullu and Mandi, and revelling in the rich pasturage. It must not be supposed, however, that this pasturage is in green meadows or downs. It is a scant herbage called "niru," growing sparsely among the rocks, and bluish-green in tint. Its quality, however, more than compensates for its lack of quantity, for sheep arriving in the country in a miserable state leave it in prime condition. The goats, however, subsist chiefly on the leaves of the birch and bush willow. Of larger trees there are absolutely none.

Shivering with cold, we admired the grand wild scenery, especially the magnificent Sonapani glacier, exactly opposite us, across the valley, and not more than six or eight miles distant. Winding through the valley three thousand feet below us lay the Chandra, a roaring river of icy water, filled with the silt brought down by constant avalanches during its earlier course, and of a most uninviting greyish-brown colour, very suggestive of goitre. It was almost impossible to imagine that the narrow strip of mountain over which we had passed could transfer us from a country densely wooded, from a smiling green and cultivated valley, to a wilderness of rock, snow, and ice-fields. But we did not linger long to ponder the fact. We were afraid of the wind; and dark icy mists came rolling up and shut out the view. Moreover, one requires very strong inducement to linger over cold meat and *choppatties*, washed down with iced whisky and water on a dark windy day, at the summit of a pass two and a half miles above the sea-level. Almost before we commenced the descent it came on to sleet. As we got lower down the sleet turned into rain, and, ere we reached Rahla, we were wet through.

In the afternoon went to visit a Lahoul encampment, a couple of miles below the bungalow, where they had some Ladakhi goats, which I had expressed anxiety to inspect. Of course, D. knew I meant buying, and it ended in my purchasing three. They were beautiful little creatures, very small and short in the leg, with very long silky hair. We became possessed of a white, a black, and a slate-coloured one. But they are not easy to keep, as during July and August they will not live at an elevation less than eight thousand feet.

We had a very unpleasant journey back to Manali, as it did not cease raining for the three following days. We heard, from the next people who crossed the pass, that there were three inches of snow on it, so we were only just in time.

CHAPTER XIV

ON the 11th of October, we marched to Ligin, a small village eight thousand feet above sea-level, to have some *chikor* shooting. It stands on the steep slope of a hill, and the fields, mere narrow terraces cut out of the hillside, only just allowed room to pitch a tent ten feet square. There was, however, one clear spot about twenty yards square, at about the angle of Holborn Hill, but we preferred to be more cramped for space and to make up for it in level ground. I strongly object to sleeping at an angle, and to eating with my chair sloping downwards and backwards from the table. So we gave up the roomy slope to the servants.

As usual, on one side of this open space, stood the village temple, beneath a stately and beautiful deodar tree of great age, and in consequence of the temple the place was considered very holy. The ideas of the people of Kullu on the subject of sanctity, are, to say the least of it, quaint. In the temple itself they allowed sheep, goats, cows, and men of even the lowest caste. But horses they would not allow, because they were shod with iron shoes! We had a couple of ponies with us. One we had just bought for three pounds. He was a Yarkundi pony that had just performed the journey from Yarkand into Kullu, over a desert of rock and snow, and up and down a dozen passes, the lowest of which was fourteen thousand feet. He was nothing but skin and bone, but we knew he would come round with feeding, so we bought him. For this journey the natives are obliged to shoe their ponies, and this one had the remains of one shoe on the off hind. On account of this piece of iron he was objected to in the temple, the only available shelter. But a donation of twopence appeased the god who presided there, and the priest in charge waived his objections.

These Yarkund ponies are not good to look at, being extravagantly ewe-necked, nor are they good movers, for they are used exclusively for carrying loads, which is not good for their action. But they are very powerful, a good height for hill use, being, on an average, about fourteen hands, and are simply perfect up and down hill and over bad roads, and are, therefore, much sought after. In crossing the desert lands between Kullu and their own country, they are frequently for one or two days at a time without food, but when they reach a place where fodder is comparatively plentiful, they are allowed to gorge themselves. Moreover, during the two-and-half to three months occupied by the journey, they are some times camped in low narrow valleys where the sun in July and August has tremendous power, and sometimes on the top of passes sixteen thousand feet high, where they are exposed to a terrific wind and frequent snowstorms. As they are entirely unprovided with any clothing, their condition on reaching Kullu is not to be wondered at.

Having got our ponies stabled in the temple, another difficulty now arose. The servants were encamped on the green slopes in front of the residence of the local divinity, but the weather was very bad, and they came and complained to us that the priest would not allow them to dig a trench round their tents, because they would dig with an iron implement! This the god would not permit. We had to administer another two pence and the people cut pointed sticks and made the trenches for them.

Not many days after, the man who looked after the ponies came to me and begged some old news papers. I asked no questions, but gave them, wondering much what he required them for. This was shortly afterwards explained, for when I went out I found him grooming his charge with bits of papers rolled up into a ball, in the shape of a wisp of hay! English grooms, make a note.

Ligin was an out-of-the-way place, where the people had few opportunities of observing "sahibs," and consequently our camp was a great attraction, especially to the ladies, who frequently came round our tents to have a stare at us. We, also, made a study of them, and their dress and jewellery.

Two days after our arrival at Ligin, the weather cleared and we had a day's shooting. Game is seldom found in the middle of the day, and as walking after *chikor* is very hard work, we rested between ten and four. We found plenty of birds, and the day's bag was ten and a half brace; nothing, of course, for England, where the birds are preserved, but a good bag for this country, and one necessitating more exercise than one would get in three days among the turnips at home. D. said that it was entirely on account of my marking down, that we did so well. *Chikor*, when flushed, always fly out of sight of the gun before they settle again. So we were obliged to post some one in a favoured spot below (they always fly down hill), where a good view of the country round about could be obtained, and of all the people we tried not one could mark. They had not the vestige of an idea that it was necessary to know exactly where the birds had gone, and we could never, never impress it on them. They would swing their arm round from north to south, from east to west, and say, "They have gone in that direction!" describing an area of three or four square miles, and an ascent and descent of several thousand feet. Whereas, if I was on the look-out and the birds went my way, D. got twice as many as if they went in any other direction. *Chikor* shooting is grand sport. To my mind they are the only birds worth shooting in the hills, and snipe the same in the plains. It is true that *chikor* beat Frenchmen at running; but, if you work up hill to them, with a good spaniel to put them up, they are sure to come back to you.

We were very anxious, if possible, to get a few red bears with their autumn coats, which are then much darker than in the spring. So on October 16th we started for a *thach* or grazing spot, some three thousand feet above our camp, to see if there were any signs of them, and to find a sheltered place where we could pitch our tent. The *thaches* lie on the very top of the ridge, and are exposed in bad weather to very fierce gales, and we knew what it was to have our tents blown down or the tent-poles broken with the weight of the snow. After four hours' hard climbing we reached the first *thach*, which was on a considerable slope, and had a long belt of firs running through it at right angles to the prevailing wind. This would not do, so we sat down and breakfasted before exploring further. As it was bitterly cold with a high wind, and freezing hard, we did not stay long. On the next two or three *thaches* we visited we found undoubted signs of the bears being about, but as there was absolutely no shelter and it had now begun to snow, we were compelled to turn our faces towards camp. When we got back to the Korse *thach*, as the first one we had visited was called, it was a sheet of snow, and did not look at all an inviting spot for living under canvas. It snowed till we were half-way down, when it turned to rain, and we reached camp at dark, wet and tired. The next morning the whole mountain side was under snow, and we decided to give up the autumn bears.

After this, when the weather cleared, we went out daily, *chikor* shooting, but took matters very easily. We could have shot any number, but we only shot as many as we wanted ourselves, and to send down the valley to friends. Even as it was, we got more than we knew what to do with, and gave away many to the villagers. Our body-servants would not touch them, as they were all professed Mussulmans, and pretended they could not eat anything that had not been killed in accordance with their creed. A little while before this our bearer fell sick, and as he looked weak I offered him a leg of our mutton, the sheep having been killed by the cook, one of his own caste. He declined it, however, because it had been placed in the meat-safe, though on another shelf, with a piece of bacon! A few days later I found this very orthodox follower of Islam drinking our whisky and filling up the bottle with water, and within a few weeks we had to discharge him for drunkenness!

"He tried on, this heathen Chinee,
But he didn't get over me!"

On October 20th we visited a small spot high above our camp, close to the forest, and the highest piece of cultivated ground anywhere about, which we had calculated likely to hold some *manāl* pheasants. We were right, for we found a great many, but owing to bad management only succeeded in bagging a couple. D. had three or four coolies with him, and walked in line up the hill, he being in the centre. Some dozen *manāl* got up, one after another, at the far end of the line, but out of shot. They flew over the nearest ridge, but the coolie at the end of the line marked them down. D. went towards him to have the place pointed out, but, as usual, could only glean from him a general idea.

There were two small patches of rocky jungle, about one hundred yards apart, in the direction in which they had gone, and in one of these D. knew the birds must be. Of course the coolie swore they were in the furthest instead of the nearest, and I had the annoyance of seeing these beautiful birds get up one after the other and make for the high forest. Had he gone to the other patch he must have at least got two, right and left, as they rose at the proper interval, one after the other. There was now nothing for it but to beat the high forest. He sent in the coolies, and succeeded in getting a couple of shots, and in killing two birds as they came past him, going down hill, simply like greased lightning, and thus his temper was recovered.

The cock *manăl* is, without doubt, the most beautiful game bird in the world. So beautiful are they, with plumage varying in colour in every light, that they are very difficult to describe. The crested head changes from bronze to green, from green to blue and purple on the neck. The back is purple to white, and again to peacock-blue and green. The wings are purple, blue, and black, and the tail orange-brown.

D.'s time was fully occupied just now in training "Nell's" spaniel pups. There were plenty of birds, and he could afford to sacrifice shots for the sake of teaching the dogs. They learnt quickly. All retrieved beautifully, and worked to the hand, and in a couple of months we sold six for £16, keeping one for ourselves.

On the morning of October 26th some men came in early and reported a bear and a panther in a small *nala* near the river. D. did not believe it, especially as they had thrown a panther in, knowing how very chary "Spots" is of showing himself, and judged it most improbable that they should have discovered his temporary abode. As the *nala* in question was quite fifteen hundred feet below our camp, he thought it rather far to go upon such a doubtful chance. But I, who have twice the "go" that he has, insisted on starting at once and risking it, as I was longing for a shot at "Spots." The *nala* was a tight one, and there was not much difficulty in ascertaining, directly we reached it, that it held neither bear nor panther, so we dismissed the news-bearers without payment, but with D.'s choicest blessings in his most fluent

Hindostanee. We always gave them a reward if their news turned out to be true; and I am inclined to think that in this case a bear had been seen about, and that they had located it in this *nala* on the off-chance, throwing in the panther, which had doubtless killed recently a cow, whose carcase lay there. We spent the rest of the day *chikor* shooting.

About this time we were in some anxiety over our poor three-pound pony, which D. had christened "Budsoorat" or "Ugly," but which I called the "hair-trunk." He was suffering from itch, the result of previous neglect and exposure. We had his body bathed with "Jeye's Perfect Purifier"; but as he did not improve, we were open to try any other remedies or advice, when my syce came and told me of a very wise man who would cure him at once. As natives have been known, by accident, of course, to hit upon efficacious remedies for certain diseases, I asked what the treatment might be. The horse doctor suggested that we should shoot a *lungoor* (large, whitebearded monkey), make soup of it, and administer it to the horse! How it was proposed to do this, or whether they imagined the pony would drink it of his own free will, piping hot, I did not inquire, for we did not try the remedy.

These Kullu natives are a quaint race. One of our men came to me once and asked for three days' holiday, and what for? He explained that a crow had been flying round his wife's head, that she had fallen ill in consequence, and that the only cure was to kill a sheep or goat, throw the meat to the crows, which would then be propitiated, and his wife recover. We gave him the required leave, and he carried out the programme, and returned to say that his wife was cured, and, of course, with a firmer belief than ever in the infallibility of the rascally *fakir*, who had given him the prescription.

We were out *chikor* shooting on October 29th, walking in line with three or four coolies, when we suddenly heard a yell on the extreme left, and discovered that we had started a huge black bear with a covey of *chikor*. He broke back through the line, and one of the coolies declared he had charged him, and would have killed him had he not thrown himself down. Of this, I must confess, I saw nothing. He broke through the line with a few loud grunts, but did not go near any coolie. It is curious, but I have never yet known of a case of a native seeing a bear within one hundred yards, without swearing that the bear went for him. They imagine that it stamps them as heroes for ever. It is a pretty sight to see a bear get up when one is out partridge shooting, and one which, I have no doubt, would be much appreciated in England.

Speaking of heroes, however, I must add a story I have heard, which goes some way to show how far the Kullu people are deserving of the name.

During the recent Russian scare, a policeman spread the report among the *lambardars* and chief *zemindars* of Kullu that the *Sirkar* (Government) had heard so much of the bravery of the young men of Kullu, that it had decided to raise by conscription a Kullu regiment eight hundred strong. The people were horribly alarmed, and being about as fond of fighting as a Bengali Babu, waited on the policeman in numbers to ask him if he was quite certain that it was true. Undoubtedly it was true, he replied, as the *subadar sahib* (police officer) who was coming to enforce the conscription was his mother's brother, and the regiment as soon as raised was to be sent across the frontier to meet the Russians, who, as everyone knows, are regular devils. They asked him what they could do to avoid service. He told them that the *subadar* was a good man and a friend of the viceroy's, and that exemption from service could be obtained, only that the stamps required were very peculiar ones. They were only to be got in the viceroy's office, and the price of them varied from one rupee to four annas, according to the means of the individual who wished to avoid service. He added that he was going on leave, and that if they liked they could give him the money for the stamps, and he would go to Simla and settle everything with the *subadar*, only that they must make haste. During the next two days this scoundrel, who was really a discharged policeman hailing from a distant part of the Punjab, reaped a perfect harvest, the rupees simply rolling in. He forthwith departed, and, of course, Kullu saw him no more. After a couple of months, however, he actually wrote and told one of his victims that the *subadar sahib* had used such influence with the viceroy that he had given up his intention of raising a regiment of Kullu warriors, and to this day they believe him!

CHAPTER XV

AT the end of October we moved into winter quarters, about fifteen miles down the valley, to the small house we had rented the previous year. The march down through the forest on that bright, frosty autumnal morning was delightful. The pine, the deodar, the walnut, and the oak were rich in varied tints. On the strips of cultivation up the steep hillsides the farmers were busy gathering in their autumn crops. We passed over mountain torrents of icy water, and were feasted with a continuous view of the gigantic snow range as we regretfully passed down, instead of up the valley. After continuous downhill of eight or nine miles, we came again upon our old friend the Beas, now no longer flooded with monsoon rains, but still a noisy roaring torrent, as can well be imagined, for though only within twenty miles of its source, it had fallen nearly nine thousand feet. With such a fall its appearance in a flood is magnificent.

The river is bridged in five different places in the first fifty miles of its course, and the bridges, all of the same kind, are called *sungha* bridges. These are simply a cantilever bridge, a scientific adaptation of which, I believe, is to be found in the Forth Bridge. On either side the river massive structures of masonry are made, interlaced with cross-beams of timber. Into these, projecting arms of timber are inserted, in tiers, one above the other, inclined upwards, and the two ends are connected by a staging made with a couple of stout trees, over which is laid boarding for a road way. These bridges, beside being of great strength and durability, have the advantage of being built with materials close at hand, and without the necessity of any machinery. When built of mature deodar, they are said to last from sixty to eighty years; but they are, of course, liable to be washed away by the floods, which so frequently occur in July and August, and are often seriously damaged by them. In 1884 the bridge below Bajaora was washed entirely away, at the moment, unfortunately, when a large flock of sheep were passing over it, all of which were drowned.

Another means of crossing the river, which is nowhere fordable, is by *jhula* bridges, made of stout cables formed of ropes of birch and willow twigs. These cables are carried across the river and firmly secured to either bank. From two to three feet below these cables a third is fixed and attached to the side cables by ropes passed over them. It is, of course, only available for foot passengers, and not to be recommended for those with weak heads.

Treading the Corn

Another means of passing from one bank to the other is by floating across on the inflated skin of a buffalo. The carcass is skinned as you skin an eel, the holes caused by the eyes firmly closed with a leather lace, and the skin, well cured and kept oiled, is apparently—for I have seen them do it—blown out with ease by the owner. It is propelled by the ferryman lying across it on his stomach and paddling with his hands and feet most dexterously, while the passenger sits astride with his feet in the water, and very often, I should imagine, with his heart in his mouth. This manner of crossing, however, is preferred to the *jhula* bridge, especially by women, who get terribly frightened when half-way across and the bridge is in full swing. They have been known to stick in the middle, unable to move from fright, and it is difficult to render assistance, as the bridge is only supposed to carry one person at a time.

The house we had taken at Raison for the cold weather was situated in the middle of the valley, four thousand six hundred feet above the sea. The chief thing to recommend it was the fairly good small-game shooting in the neighbourhood. Unfortunately it lay midway between the only two post-offices in the district, and we daily had the satisfaction of seeing our mails pass our very door some eighteen hours before we could get hold of them, which was most peculiarly exasperating when the English letters were in the post-

A *Jhula* Bridge in Lahoul

bags. These were carried on men's backs, slung over a pole, with a sharp spear-head at the end, and covered with bells, the runners averaging five miles an hour, with a change every few miles.

A great disadvantage of residing in this civilized portion of the valley was the proximity in which we lived to the *loogri* shops, the temptations of which were sometimes more than our servants could withstand. *Loogri* is, as I have already mentioned, a foul and pernicious drink, made of unfermented rice, in appearance like very thin water-arrowroot. The smell of it is horrible, and the taste, I should imagine, equally detestable; but I never had the courage to try it. It closely resembles *chang*, so much drank in Thibet. Many of the people of Kullu are slaves to drunkenness, and spend all the money they can get in enriching the Lahoulis, who are the brewers of *loogri*. It is highly intoxicating, and its effect on the constitution very deleterious. These Lahoulis are mere settlers in the valley during the winter months, when their own country is under snow, and if a heavy tax were imposed on them as settlers, the liquor would diminish in quantity and increase in price, while a benefit would be conferred upon the people of Kullu. *Loogri* is imbibed in large quantities at all the *melas* or fairs, and is largely responsible for the debauchery and profligacy which prevails on these occasions.

Another of the drawbacks to Kullu as a residence for all Europeans, except those in robust health, is the absence of any European doctor within a hundred miles, while a dentist has never been seen in the country, the blacksmith replacing him with more or less success. I have known Europeans, in much suffering, resorting to him, and although I never heard of his successfully extricating a tooth, I believe the blacksmith has before now been felled to the ground for breaking it. With the natives he is more successful. If the tooth causing the pain is much decayed and, consequently, difficult to get out, the amateur dentist draws another somewhere in its vicinity, that is perfectly sound, and of which he can get a good hold. This he exhibits with much pride to his victim, who imagines one tooth as good, or as bad, as another, and forthwith pays the fee, a few pounds of Indian corn.

Fortunately, during our stay we did not require the services of this functionary. In this fine, dry climate, moreover, there is but little sickness. Fever is prevalent during the autumn months, when the people are at work in the wet rice crops, but it is of a mild type.

CHAPTER XVI

WE had received an invitation to spend Christmas at Bajaora, about twenty miles further down the valley, on an estate belonging to Colonel Rennick, a well-known soldier, who commanded a brigade in Burmah. He was at the time away, serving in Assam, but his wife and a friend, Miss R, were wintering in Kullu.

We started on December 24th, on a beautiful but cold morning, with nine coolies carrying our portmanteaus and bedding. For it is the custom in India to take your bedding wherever you go to stay, be it hotel or private house. In the old days, *sahibs* used even to take their own knives and forks and cruets out to dinner-parties; but this latter custom is obsolete, except when in camp. On the way to Bajaora there passed us, as we were sitting near the roadside eating our tiffin, a native nobleman on his way down country to bathe in, and drink of, the sacred waters of the Ganges. He and his retinue formed a curious procession. The ladies of his harem, ten or twelve in number, rode, but his wife was carried in a *doolie*. Behind them came a host of followers, some wearing the royal scarlet, and all filthy dirty, and then a number of men carrying bear and leopard skins stuffed with straw; this, I imagine, by way of showing what a mighty hunter he was, though most of the skins were probably presents. It is the custom for natives thus to treat their trophies. A skin, nicely cured and thrown over a couch, or on the floor, is not sufficiently ostentatious for them. Skins treated as they treat them never last long; they are cured and stuffed in the rudest manner possible, and in a few months insects get at them, and the hair slips. They care not, however, for, child-like, a native only cares for a thing so long as it is novel.

We reached our destination as the evening was closing in, and were astonished to find such a charming house in the valley, so well planned and built, and capitally furnished, even to electric bells! It was, in fact, a gentleman's house, its only fault being that it was too good for Kullu. The owner, a great sportsman, intends either to sell his entire property for a lac of rupees, a little over £6500, at the present rate of exchange, or to end his days there. He has lately built another house, twenty-five miles higher up the valley, at Nugger, and owns many thousand acres in Kullu, so would doubtless find plenty to occupy him on his estate, though less than on a property the same size at home, for the land is all handed over to tenants who work it, paying the landlord by half produce, and the latter paying the revenue to Government.

Ploughing, Lower Kulu

The outhouses, the cattle-sheds, the stables, are all built in the same excellent way as the house, no expense having been spared. The gardens are well laid out, and full of the best kind of roses, and of many choice English flowers; the tennis-court is a good one, and the orchards planted with hundreds of trees—apple, peaches, pears, apricots, oranges, lemons, almonds, plums, and pistachionuts, while a constant flow of water, which can be turned on in any direction, renders the prosperity of the gardens independent of the rainfall.

It is altogether a charming property, and absurdly cheap at the price Colonel R. is asking for it. At Simla, the hill capital of India, only about one hundred and twenty miles to the south-west, there is a ready sale for any amount of fruit, and the price paid for the apples and pears, for which Kullu is renowned throughout India, is so good, that by this means alone a certain income is assured. For sport, the houses are admirably situated. At Bajaora, black partridge is found in every direction, and the hillsides are well stocked with *chikor*. The Kundi valley, at the mouth of which the house stands, is renowned for its pheasant and wood-partridge shooting. *Goral* and panthers can be shot all the year round, and bears in June, July, and August. On the opposite bank of the river, and about three miles higher up, is the entrance to the famous Parbatti *nala*, one of the finest, if not the finest shooting *nala* in Kullu, celebrated for its red bears, ibex, and *burrhel*. The other house is no less favourably situated for sport up the valley at Nugger, where any number of *manāl* and argus pheasant can be bagged during the winter months. The view from Bajaora is beautiful beyond description; it includes, during nine months of the year, the snows of higher ranges in Kullu towards the Rohtung Pass, and the lofty mountains of Lahoul. For those who like a quiet life, with good sport, a perfect climate, where epidemics are little known, and the scenery equal to any in Kashmir, here is a chance indeed. I rather fancy the colonel would let his house on lease, as he has still some years of soldiering to put in.

From the house at Bajaora we had a capital view of Bijli hill, so called because the temple at the top, the Bijli Mahadeo, is so constantly struck by lightning. The hill rises four thousand feet almost perpendicularly from the river, and although thunderstorms are of very rare occurence in other parts of Kullu, it appears to possess a peculiar attraction for thunder-clouds, and so, inspite of the rams' horns with which it is profusely decorated, and which the natives firmly believe will ward off any stray flashes, it is constantly struck by lightning.

At the foot of this hill and close to the junction of the Beas and Parbatti rivers, is the burning *ghāt* for members of the Kullu royal family. The late rajah's father was burnt here some four years ago. He lost his life in a remarkable way. It seems that he was visiting some Kullu ladies and was amusing himself with them in the upper story of their house, when he became too demonstrative in his attentions, and they forthwith and promptly, chucked him out of the window. Being old and excessively fat (in accordance with the sealed pattern for rajahs) this proceeding was too much for him, and he was picked up dead.

The rajah who succeeded him (or rather *rai*, for the old title of *rajah* is not recognized by the Indian Government in the case of Kullu) is a dapper little man, who speaks, reads, and writes English well. But he is rapidly attaining the royal figure, and I expect in a few years will be as round as a ball. He shoots a little, but his great pride is his harem. He has fourteen or fifteen wives, and as he spends more than he gets on these luxuries, he is more or less indebted to a rich banker in the town of Sultanpore. This banker was one of those lucky people who made his pile in the purchase of sapphires.

Incredible as it may sound, only a few years ago these precious gems were being hawked about Kullu and sold for a few measures of grain. No one knew their value, and they were sold as "blue flints"! The European residents in Kullu actually turned away from their doors people offering them basketfuls of these almost priceless jewels for a few pence. At last it was discovered that they were sapphires, and amongst those who benefited by the *éclaircissement* was the banker, who having purchased a few basketfuls, found himself a millionaire, having previously been a poor mule-driver. Immediately that it was discovered that the "blue flints" were sapphires, the mines, which are in Ladakh, and belong to the Maharajah of Kashmir, were closed and carefully guarded, and now but few of the precious gems find their way into Kullu.

Occasionally, however, one meets natives offering some for sale, but they have almost certainly been stolen from the mines; and, as their value is now perfectly understood, there is no chance of picking up bargains. I need hardly add that, to this day, a bitter wail ascends from the entire valley of Kullu—*zemindar*, planter, official, and missionary alike being ever haunted by the recollection that they have thrown away a chance which the goddess Fortune had once placed within their grasp, and which will never come again.

PART-FOUR

UNDER CANVAS IN KULLU AGAIN

CHAPTER XVII

ON the 1st of April, 1892, we set out once more for our happy hunting-ground. How very little we poor mortals know about the laws which govern the weather! The previous monsoon had been practically a failure everywhere. There had been a good fall of snow on the Himalayas on September 30th, and the knowing ones predicted an early and severe winter. What really happened was, that snow fell but three or four times throughout the winter in Kullu, and that the summer set in, in March, fully two months before the usual time. No living inhabitant can remember such a mild winter, and it has been a disastrous one for the agriculturists of the valley, who are dependent on the melting snow, not only for the irrigation of their crops, but also for drinking purposes.

But though the country was more free from snow now then even in the previous July, we could see no signs of bears. The hibernation of these animals has always been a standing puzzle to me. I used to be under the impression that, when food began to get scarce in autumn, they ascended the mountains and made themselves comfortable under a suitable rock or in the hollow of a tree, where they slept covered by the first fall of snow, and that there they remained till they awoke with the melting of the snow in the spring, and immediately commenced to make up for lost time in their search for food. My theory, however, now received a rude shock. Even up to twelve thousand feet there was not a flake of snow to be seen, except here and there in deep shade. On April 6th, up to eighteen thousand feet there were but a few inches of snow, and yet not a sign of a bear could be found in all the country round. The previous year, when there had been an abnormal snowfall, they began to emerge from their winter-quarters early in May, but now it would seem as though they did not intend to show themselves till the same time, though there was no snow. What, then, is the law which governs their hibernation? Do they sleep for a certain number of days, or is their sleep influenced by the early or late advent of the warm weather? Now it so often happens that though there is no snow, the intense cold of the nights has retarded the growth of the grass, and the grubs to which the bears are so partial, and which they find under the large stones on the different *thaches*, are all burnt up by the heat of the midday sun. We began to hope that perhaps when the grass had grown sufficiently to give them a good square meal, the bears would show. But, in the meantime, where were they?

We selected the Solung *nala* again as our quarters, because, in spite of the number of bears we got out of it last year, we believed it to be the best in Kullu. It is many miles in length, and there are eight or ten smaller *nalas* running into it, all likely to hold game, and it has, moreover, the advantage of being more difficult of access than any other. Near the mouth of the *nala* lies the village which bears its name. It is remarkable for the possession of the largest deodar tree to be found in India. We measured it one day our selves, and at six feet from the ground found it to be thirty-eight feet three inches round, and that it carried the same measurement up to about forty feet from the ground!

Our camp was far above the village, and, except for getting our daily supply of milk from it, we did not trouble Solung. At first we were hardly comfortable under canvas. Down came the rain, the wind blew, and the snow fell too, down close to our camp. When encamped at a great elevation, foul weather is simply abominable. The wind howls through the tent; it is so dark that it is impossible to see to read, and equally impossible to keep warm.

On the 7th of April, after a pouring and blowing day, we lighted the lamps at two p.m., and read by them till we were tired, and then, for want of something better to do, went to bed at half-past seven! The temperature in the tent was 38° Fahr. We complained, indeed, often, of the want of change in our food; we had no need to complain of a want of change in the temperature. D. noted it as follows, on April 4th, in the tent: 7 a.m., 35°; 8.30 a.m., 80° (dead calm); 12 noon, 68°;* 4 p.m., 50°; 9 p.m., 40°. Like all tents in India, ours were, of course, double tents, as a protection both against the in tense heat in the middle of the day and the great cold at night. But on a cloudy day they were extremely dark, and it is unnatural and depressing to have to read and write by lamplight during the brightest part of the day. It is the lesser of two evils, however. When D. was in Cyprus, with the army of occupation, in 1878, he became convinced that the cause of the great amount of fever, common among the troops, was the fact that Government had only provided "bell-tents" instead of India double-fly tents. The difference of temperature there, during the twenty-four hours, was enormous.

This unpleasant change, all in one day, from summer to winter, was explained to us soon, in a note we received from a resident down the valley. It appears that the people of his village feared a famine on account of the dry, hot, and unseasonable weather. As their prayers remained, apparently, unheeded, they took the matter into their own hands. Dragging their gods out of the temples, they locked them all up together, and heaped every kind of abuse upon them. The gods relented, and the change I have described really did take place; moreover, snow was prophesied for May 1st, the day of their annual great *mela*. We had not much fear, however, of the prophecy coming true!

In spite of cloudy and threatening weather, we made our way next afternoon to Look-out Hill, passing on our way through a magnificent walnut forest. The trees were, of course, leafless, but the ground beneath them was literally strewn with walnuts. The villagers use what they can, grinding them up for walnut oil to cook with, but there is no means of transport for what, once in India, would command a good sale.

Next day was beautiful, but piercingly cold. I tried to warm myself at the camp-fire we keep up, chiefly for the benefit of our camp-followers; but I

---

* This was after the day breeze had sprung up.

consider camp-fires a fraud. One is alternately cooked and frozen, as one stands back to or facing it. During the abnormal weather of the few days just past, we had been obliged to build a shelter for our ponies, composed of four uprights, roofed with fir branches. Over this we piled all the hay we had collected from a village five miles below. This made a capital protection from the wet, and one which we had found useful for our retainers before, when marching in Kashmir. But now, in Kullu, we had six tents, and so they needed no other shelter. In these fir-branch linheys we now left the ponies for some weeks, as it was impossible to take them further up the *nala*.

CHAPTER XVIII

AS far as Solung we had brought a cow. But fancying that she would suffer from the cold, we had her driven down the valley again. We now heard that she was dead, from injuries received on the road. Two coolies had been intrusted to take her down, and No. 1 declared that while he was stopping behind to eat a mouthful, coolie No. 2 allowed her to fall over the precipice. No. 2, however, accused No. 1 of having beaten and kicked the cow unmercifully *en route*, quite enough to kill her. We believed the latter story, as coolie No. 1 had been some time in our service, and D. said he would back him, in the matter of lying, against any other native in India, which sounds a "tall order."

But, as we were unacquainted with the Kullu language, we sent the two worthies to be examined by a friend, who, unable to arrive at the truth, proposed to ascertain it by the cheap and speedy method of the goat-test, before alluded to, administered by the Pundit of the neighbouring temple. In this case, however, the proceeding varied somewhat from that at Nugger. The Pundit, in the presence of the idol of the temple, who was appealed to, would sprinkle the goats, and then let them loose. Whichever goat walks first into the temple, his owner is adjudged the veracious one. If both walk, both owners are to be believed; if neither goat enters, both the coolies are liars. The people prefer a trial of this kind to one before any living judge. If I lived near the temple I would take care to keep a goat which I would feed daily inside, and then go in heavily for litigation!

Accordingly, the trial came off solemnly in the presence of the priest and some forty spectators, and the *devta* made the goat of coolie No. 1—him in my service—walk into the temple, while that belonging to coolie No. 2 fell to the ground. The blame of the cow's death thus fell on the latter's shoulders. But we did not agree to the verdict, much to the natives' surprise. There was a path practicable for animals, and another, a shortcut, impracticable for them, on account of the one dangerous spot at which, as was shown by blood on the road, the cow fell. As our coolie should on no account have permitted her to take that road, we held him responsible, and discharged him.

Verily the people of this country will eat anything. Grasses, jungle weeds, fungi—all they devour. I now made the discovery that they will eat horse-chestnuts, which, till now, I imagined to be poisonous. They gather them and put them in holes about a foot deep to dry. Then they are broken into small pieces with a wooden mallet, washed in several waters, and when quite clean, left to soak for several days. They are then taken out, dried, and ground into flour, and accounted delicious. Why does not someone at home try "chestnut flour"?

D.'s rifle-coolie of last year came up to us on April 12th. He had been wintering for the sake of pasture in the native state of Mandi, with his two hundred sheep, and brought alarming accounts of the terrific heat down country. He declared that the very stones were on fire, and that many of his sheep had died on the way, and the remainder had been half-starved. We took him on again as rifle-coolie, because he knew his way about these mountains, which we had found, from experience, to be a consideration when overtaken by darkness.

We now tried a *nala* which we had yet never visited. Though we followed it up to a narrow gorge choked with snow, where we could proceed no further, we saw no living animal, but the *nala* positively swarmed with beautiful *manāl* pheasants. It is a very steep *nala*, and evidently annually devastated by avalanches. The havoc these had wrought among a magnificent forest of deodars at its mouth was quite pitiable. Hundreds of these fine trees lay heaped one on the top of another, reminding one of a giant's game of spilicans—but spilicans some two or three hundred feet long and of enormous girth. It was quite sad to see, this forest in ruins, this abomination of desolation.

The continued hot weather not only burnt up all the bears' food, but also made stalking difficult if not impossible. It was not at all easy to keep one's feet on the steep mountain-side, and as there been no snow to decay the fallen leaves, they lay as they fell in the autumn, crisp and dry, and cracked like a pistol-shot beneath one's feet. In addition, the continued heat brought a plague of flies, especially blue-bottles, and the "ticks" swarmed in millions. If they were bad last year, this season we found them simply a veritable plague. There was no avoiding them, and I was constantly covered with them, and suffered a martyrdom. The natives say their bite brings fever, and I must say it is exceedingly poisonous, and leaves its mark for two or three days. D., whose costume was more adapted to protect him, suffered less. I have since found a remedy.

About this time a new post-office was opened in the valley, at a place called Jagatsukh. It would have been a great advantage to us could the postmaster have spoken or even read English; but as it was, all our letters and papers had to be redirected in the vernacular at the nearest place where English was understood. We sent down to try and get some stamps, forms, etc., but the Babu did not understand what we required, and our messenger returned without them. Later on we received some letters and papers of ancient date, sent with a note by a resident in the valley, who wrote: "The post-office Babu was here a few days ago, and gave me the accompanying letters, etc., for you. As I expected you to send down shortly for some butter, I took them; but as you have made no sign, I now forward them to you by a coolie. The Babu wanted to give me a registered letter for you, but this responsibility I declined to accept. So he took it away with him."

Such are some of the peculiarities of her Majesty's postal service in these parts, and it will be seen the new office was a doubtful blessing. But the manners and customs of post-office officials are curious in other less remote parts of India. Natives are inquisitive to a degree, and post-cards should never be used except for very trivial matters. Some time ago our bankers notified to us that they had forwarded us a cheque book. A fortnight later, as it had not been delivered, D. wired to the bank to that effect. Before the reply came he received the cheque book. The telegram had passed through the office, where it was detained, had been read, and a search made for the missing book. It was simple, if unbusiness-like. Think of the correspondence we should have had with the General Post Office in England!

Again, a short time since the postmaster of a certain place went to a *sahib* living close by, and told him that that morning a Mr. — had sent a registered letter through the post, which he believed had reference to a matter in dispute between the two! On another occasion a postmaster in Kullu wrote to a resident who owned a dog with pups, and asked him to lend him two hundred rupees, and to make him a present of a dog child!

Further, in India, telegrams are by no means inviolable. A few years ago, when we were in Kashmir, D. received the offer of an appointment, and wired his acceptance of it. Leaving Kashmir, after a journey down country of nearly a thousand miles, he stopped at A— to change lines. As he was waiting on the platform, the stationmaster came up and condoled with him on not having got that appointment! What appointment? D. naturally asked. The one at M—, the station-master made answer; and then it transpired that after D. had left Kashmir a telegram had been sent him that he could not have the appointment after all. This message passed through A—, where this station-master occupied his leisure moments in reading the telegrams. Having ascertained who D. was, by the name on the latter's luggage, he lost no time in communicating the intelligence, to the intense surprise of the recipient.

As a fitting pendant to these true Anglo-Indian yarns, I must add another little story, written to me, from a friend at Dharmsala, and which enlivened our somewhat gloomy spirits, drooping from the non-appearance of the bears. My friend had a European maid, who, like all other European maids in India, had many aspirants to her hand among the crowd of Thomas Atkinses. One day she came to her mistress and asked permission to get married. The latter inquired who was the favoured one. The girl replied he was a soldier. "Really," replied the lady. “And what’s his name?” “Oh, ma’am,” replied the bride-elect, “I don’t know him well enough to ask him that yet!”

CHAPTER XIX

FOR fifteen mortal days we had toiled over every likely looking spot of this large tract of country without coming across the sign of a bear. However, on April 17th, we climbed a narrow and remote *nala*, not far from our camp, indeed, but so insignificant that we had hitherto neglected it. At about eighteen hundred feet above the camp we came upon unmistakable signs of fresh damp digging, not many hours old. We therefore proceeded with great caution, expecting at every ridge we approached to find the bear feeding on the other side of it, when all of a sudden D. slipped, and in so doing, unfortunately dislodged a big rock and set it rolling down the hill. The rifle-coolie tried to stop it, but in vain, and with a crash, bang, and a roar, it bounded down fully four hundred feet into the stream below, and the hills around echoed and re-echoed with the noise of its fall. Just our luck! No chance of a shot again that day.

Next morning a coolie came in to report a large black bear in a *nala* about a day's march from where we were encamped. As these people have no idea of distance, a day's march might mean anything, but we calculated it to be about ten miles. The Kullu folk estimate distance by "ek tamāku," one tobacco; i.e. the time it takes to smoke one *hookah*, and this they call three miles. As it seemed a long distance to go on the chance of one black bear, we sent one of our own men back with the coolie to watch that evening and report further. He returned next morning, saying he had again seen the black bear, and also, on a *thach* above, an enormous red bear. So we decided to take one tent and go up for a couple of nights.

When we reached the place, however, luck seemed still against us. No bears were to be seen, and though D. came across a *kustoorah* in the evening, he was in the jungle, where stalking was impossible, for the dry leaves made as much noise under our feet as the shingle on the Brighton beach. Our camp, too, was most uncomfortable. It was on a slope of about one in nine, and so slippery, that without Alpine hob-nailed boots a foothold was impossible. We upset water, milk, and rangoon oil; one leg of each of our beds rested on a stone six inches from the ground, and our table was so rickety we hardly dared look at it. In fact, the spot was so miserable that, bears or no bears, we evacuated it by daylight next day.

There was a little excitement in camp, however, before we left. We had bought a cock and three hens from a resident, and looked forward to an occasional fresh egg for breakfast. For the first few days our hopes were realized, and then the supply suddenly ceased. As a hen was still to be heard clucking regularly, we suspected that the eggs were being stolen. Now amongst our retainers we included, by necessity—for some classes of servants are exceedingly difficult to secure for a nomad life like that we were leading—a gaolbird! So, acting on the principle of setting "a thief to catch a thief," D. sent for this individual and explained matters. Within twenty-four hours he came to me, saying, "I have found the thief. If the *sahib* will go to the tent and look under the bed of one Bukroo, his High Mightiness will find an egg, which I have just seen him take and hide!"

D. forthwith descended upon Bukroo, and ordered him to pick up the roll of evil-looking *resais* and blankets, he called his bed. Perforce he had to obey, but exhibited the greatest astonishment when the egg was found beneath it, and even went so far as to suggest the hen had gone there to lay! This was adding impudence to theft, and D. promptly thrashed him till he, D., was sore, fining him a rupee, and offering him another instalment at the same price. But Bukroo did not think fit to avail himself of the offer.

Thieving counts for nothing in this country. Half an hour after this occurrence the thief and the informant were sitting amicably smoking together in the same tent!

We encamped in the Chenag nala, and spent a fruitless morning next day, going over several *thaches*. In the afternoon we went out again and spent an hour seeking a suitable spot whence to reconnoitre. Hardly had we sat down to do so when I spotted a bear far, far above us. D. looked at his watch and calculated that it was just possible for him to reach him before dark, and so set off at once. After a terrible climb up the bed of a mountain torrent, he reached the place at 6.30 p.m., and on looking round could see nothing of the bear. He turned to look for me, whom he had left far below, and, at last, with his glasses made me out, still sitting in the same place. I, meanwhile, had tied my handkerchief to my alpenstock and was holding it over my head, inclined to the right. D. perceived in an instant that the bear had gone on up, and to the right, and though he was very beat, set off in that direction as fast as he could. Before he had gone far he peeped over a slight rise and saw the bear a hundred and twenty yards off, looking straight at him. To fire was impossible, as D. could only just see over the ridge. To move would have been to make him bolt instantly. So during the space of more than five minutes he remained stock still, and the bear went on looking at him. At last the latter appeared satisfied that no one was about, and lay down to sleep. Unfortunately, however, there was no daylight to spare in order to allow him to go well off, and for D. to get a chance of approaching nearer to him. In a few minutes it would be dark, and D. was compelled to move a step or two in order to get his rifle over the ridge. In so doing of course the dry reeds on which he was standing cracked loudly. Up jumped the bear, and D. fired. But his cartridges were last year's, and must have got damp and been dried again, for the smoke hung round so after he had fired that he could see nothing. When it cleared away he ran up to the top of the ridge, and there he saw his friend, standing looking at him.

D. had no excuse now, but—he fired and missed!

It was a horrible, shameful miss; he saw the bear no more, and thus lost our first chance of the season. He abused himself pretty freely, he says, and felt almost too ashamed to come back to me. But a night spent at that altitude and without a blanket would hardly have been enjoyable, and he started to return to camp as quickly as possible.

The ascent had taken him two hours and a half, and he calculated it would take him a hour and a half to descend even by daylight, and it was quite dark now, and the road dangerous. I from below had seen him fire, and when I heard the second shot hoped it was all right, so I turned to hurry home and send a lantern to meet him. But, I too, got benighted, and had to be hauled and dragged by my evil-smelling rifle-coolie over all sorts of obstacles nearly the whole way. It was a most unpleasant performance, but I was utterly dependent on his help for getting back at all. He wanted to carry me on his back, but that I declined.

Fortunately D. had taken with him a man from the nearest village, and without him would never have found his way back at all. Even this fellow, who knew the mountain well, had to scramble along a great deal on his hands and knees. It was ticklish work, but, by following his guide closely, D. eventually reached the *thach* where he had left me, and far, far below saw the glimmer of the lantern advancing towards him. In less than half an hour he met it, and reached the camp without further difficulty. But words cannot paint my disappointment when I discovered that he had returned without the bear; as, when I heard the second shot, I felt certain that he had bagged it.

CHAPTER XX

OUR camping-ground was so uncomfortable that we really could not remain there any longer, though it was an exceedingly beautiful *nala*. We had been here before in the March of the previous year, but never so high up, and had no idea of its beauties. This Chenag *nala* is a good one for sport, as it holds *kurt* (or *tahr*), *goral*, and musk-deer, as well as bears, and there are some good camping-grounds in it. The view from the Mundran *thach*, near the head of the ravine, is beyond description grand. Exactly opposite, though many miles off, is the zigzag road leading up the Rohtung Pass, the summit of which is visible, and beyond it the glaciers of the Sonapani *nala*, in Lahoul. Below, lies the fertile Kullu valley with its beautiful deodar forests and its green fields, and to the right the Hamta Pass, fourteen thousand feet, with the snowy peaks of Spiti towering beyond it. A finer view cannot be found in Kullu, and I suspect it even surpasses the famous prospect from Malana, of which we had heard much.

On the 21st of April we came to the conclusion that the people must have been abusing their gods again, as it rained heavily all night, a most refreshing change. About midday the barking of our dogs announced the approach of a visitor, who turned out to be Mr. M—, an old forest-officer, who lives in Kullu, and who was on his way to Lahoul and had come to call. He was wet through, but a glass of whisky soon dried him, for he was as hard as nails and nothing would be likely to do him any harm. An ardent sportsman, he is accustomed to roughing it, and carries, I think, his spartan habits unnecessarily far, for he never even takes a camp bed with him, or even a light, or many other things which would add greatly to his comfort without adding materially to the expense of transport. But *chacun à son goût*.

After he left us on his way to Rahla, at the foot of the pass, we went out to look for a bear, and walked till dark, coming across some old marks, but saw nothing. At one time we thought we were near a bear, and D. laid it upon his rifle-coolie to take extra precautions not to make a noise. By the way, are Europeans the only race that can whisper? It is a most difficult thing to teach a native the art, or trick, or what ever it is. Yet it is essential when big-game shooting. But they learn with difficulty, and always want to whisper through their throats, which makes as much noise as talking aloud.

The Hindoo Lent, we were glad to find, was now nearly over. During these twelve days, in Kullu, at all events, the Hindoos appear debarred from cutting wood, or using iron in any way, though there would appear to be no restriction on eating or drinking. Our rifle-coolies, too, had been perpetually asking for leave. So we were by no means sorry when the season came to an end.

We had shifted our sleeping tent a few yards, near a tree with some withered branches at the top. As I feared that these might fall on the tent, we sent a low-caste native up the tree to lop them. As he could not swarm it, we made him use our iron tent-pegs for steps, and gave him a long rope to attach to the branches, that D. might drag them away from the tent as he cut them off. The man got up without difficulty; but, unfortunately, in so doing broke his coral necklace. A baby could not have been more distressed. These people nearly all wear coral necklaces of common, deep-red beads. Turquoise they rather prefer, when they can get them, but the turquoises they wear must always have flaws in them to prove their genuineness.

The coolie having climbed to the branches and pronounced them safe, D. told him to come down and leave them. But, before he descended, he let down the rope to the bottom of the tree, measured the exact mark from where he stood to the ground, tied a knot in the rope to mark it, and then, having gone through a form of prayer, began the descent. I asked him the meaning of this performance, and he replied that had he not done so he would certainly have fallen and been killed. I wonder if he prayed for the recovery of his coral beads!

One cloudy, windy morning, April 23rd, we decided to see if, perchance, the bears were high up, just under the snow, and to investigate the Putsaru *nala*. Taking our tiffin with us, we made an early start. Rain and sleet came on soon, and when, about one o'clock, we sat down for a bite, we were speedily glad to get under way again. It cleared a little and we enjoyed the walk, following the course of the mountain stream. The banks on either side had been levelled by avalanches, and the walking was not difficult. It was, as D. said, a pretty *nala* for a bear: on either side forests of silver birch, with green banks sloping to the torrent. Every moment we hoped to come across one, but in vain. At last we emerged on a large rock-covered plain, brown with the short grass of last year, as it was too cold for any fresh grass to have yet sprouted. We wandered about without seeing a sign of any living thing. A gentle slope of about a quarter of a mile would have taken us to the crest of the mountain, whence we should have had a grand view into Lahoul. But we were afraid to move on, as on looking back we saw heavy black clouds, solid as a wall, rolling up the valley towards us, and knew that we were in for a snow storm. So we continued along the same level, intending to descend by another *nala*. But the snow soon came upon us in great heavy flakes, and it grew so dark that there was no chance of seeing a bear, even should one venture out in such weather. For bears are very careful of their coats, though why, I am at a loss to understand. No rain could ever get through them, but probably they would become so heavy when sodden that the animals would be considerably handicapped if called upon to run for their lives. Much to my disgust—as I thought we were giving up any chance of seeing game, but both D. and the rifle coolies were against me—we descended by another and shorter *nala*, than that I had chosen. The snow changed to sleet about fifteen hundred feet lower down, and we reached the camp drenched to the skin, but satisfied that the bears were not to be found high up.

For some time I had been missing potatoes. This may sound a trivial matter to any housekeeper who can send round the corner to the greengrocer and buy more. But here, in Kullu, where they are very expensive and only to be got with great difficulty, it was no joke. When they were gone, we should be left without any vegetable, as I had not been able even to purchase an onion for three months. Every week I sent down into Sultanpore for some, and every week came the answer back that there were none to be had. So, when I found my potatoes vanishing, I again called in the services of my private detective, whom we had now nicknamed "Polaky." In a few hours he returned with the information that my cook was the delinquent, and that the missing potatoes were to be found in a basket in his tent, but that I must look sharp, or he would cook them. D. went and investigated, and there, where I had been told, found a large bag of potatoes, some pounds of our meat, and some of our flour.

Now servants in India feed themselves, and have no right to any of their master's comestibles. Had we searched further, we might have made more discoveries. But when the cook came back to his tent, we informed him that we had sent for a coolie to carry his baggage, and gave him one hour to clear out. And he went.

CHAPTER XXI

THE following evening we at last saw a bear. I spied him feeding, but unfortunately D. could not get at him without going back a long way, as the only bridge over the river was the tree-bridge we had made. While we hesitated to see which way he was going, it came on to pour. Br'er Bar, careful, as usual, of his ulster, vanished into the jungle, and D. did not follow him. We remained sitting in the rain for more than an hour, fearfully cold. Directly it cleared, out walked the bear on to the thach about a couple of hundred feet above the place whence he had vanished to take shelter. He was a magnificent fellow, but it was now too late to go after him. The man who hesitates loses his shot. D. fancied he was the same dark-coloured bear he had seen last year, and whose skin he coveted exceedingly. So we laid deep plans for the morrow and returned to camp.

We went out at four a.m., I to the spot whence I had spied the bear, and D. crossed the river, and made direct for the place where he had been feeding. It was a lovely morning, and freezing hard. The sun first touched the snow peaks twelve thousand feet above us, at a quarter to six, and very lovely they looked. There is some thing most fascinating in watching the sunrise as it slowly creeps down a snow peak, turning it all rosy. But we were in no luck that day. I, nearly frozen, watched every likely spot from my coign of vantage, and D. investigated every thach within two miles of that on which the bear had been greedily devouring large quantities of the common stinging-nettle, for which bears have a taste in common with the Kullu natives.

We sought that bear again in the evening, searching every open space in the jungle with our glasses. Just before dark I spotted a bear across the river, but there was no chance of getting at him. The next day we went after him, but he, too, had vanished. During our walk we crossed a *nala* high up, and had some difficulty in getting over the stream, which was in flood. Later in the day, when returning across the same *nala*, but lower down, imagine our surprise at finding the bed of the torrent perfectly dry! At first I thought we had missed our way, and that it was not the same *nala*. But we found that the stream had suddenly disappeared into the ground about a quarter of a mile further up. This was the more astonishing as the dry bed showed traces of having been recently covered with water.

A few days of very unsettled weather now followed, evidently the result of a succession of dust-storms in the plains. Though we were more than two hundred miles as the crow flies from them, and more than a mile and a half (nine thousand feet) above their level, the effect of one of these dust-storms was very apparent. The under-sides of the leaves on the trees were of a fresh delicate green, while the upper sides were thickly coated with dust. Even the snow capped mountains far above us, up to twenty-two thousand feet, were brown instead of white from the same cause. Last year we had none of these storms, but this season we experienced several.

The dust even got into the milk—according to the villagers. I remonstrated with the man, who brought the milk from the nearest village each morning, on its dirty appearance. He explained that there was so much dust on the grass the cows ate that it affected the milk. And he believed it, too!

Twenty-four hours' steady rain set in, but even after that the dust-storm did not abate. The dust came sweeping up the valley in great clouds, smothering everything, and almost blinding one, it was so fine. It lasted three days. Heavy rain and tempest followed, and we were confined to our tents all one day, very dull and very cold. Books were scarce. We had devoured all our last English mail, our weeks-old news papers, and we were weary of making plans to circumvent imaginary bears.

A most unpleasant "divarsh'on" was the killing of a sheep in camp, but on these occasions I always went far away. As usual, all our retainers came flocking and begging for pieces. Except our body-servants, who were all

Mussulmans, none objected to any portion of the animal, nor had they scruples against eating bear's flesh. I wondered if I should make them a handsome present of two dozen eggs I had sent for down to the village, and which turned out to be all rotten.

A fine morning following the rain, we paid a visit to Dhundi, to see if any bears had been at work on those beautiful *thaches*. But we found not a sign. We went up the Palchan *nala*, towards the glaciers, but found not a trace. It was becoming serious. Excepting at a *kustoorah*, which D. bagged, we had not fired a shot during the twenty-eight days we had been in the *nala*. While walking up to Dhundi we put up several woodcock, which confirms D.'s theory that these birds breed in the valley, and are not birds of passage, as in Europe.

To the plague of dust succeeded a plague of locusts, evidently driven up from below by the storm. Last year they were about the valley in myriads. They get driven into the snows by the wind, and there perish of cold. An officer on survey duty in Bara-Bagahal last autumn, wrote as follows:—

"The natives say that the bears have gone mad from eating locusts, and certainly they have not far to go to seek this food. The hillsides are literally covered with tons of dead locusts, and the stench in some of the upper *nalas* from the rotting insects is simply awful!"

A greater scourge, however, than the dust or the locust, had invaded Kullu. On the 1st of May we received the alarming intelligence that, for the first time for forty years, cholera had broken out in the valley. It was said to have been brought in by two policemen sent from the district of Dharmsala, where it was raging. One man died *en route*, and the other came on, bringing, probably, the dead man's clothes with him. He, too, died at Sultanpore immediately; and thus, by a single act of rashness on the part of the authorities, Kullu was laid open to the plague, and with what dire results I shall later have to relate.

As I, personally, had had a very unpleasant experience of cholera when in Kashmir a few years back, we decided not to run any risk, but to march on up into Lahoul, where, on account of the great altitude and extreme cold, we might hope to enjoy an immunity from the scourge. Whilst on the subject of cholera, I may mention it has been noticed for many years, not only in India, but also in Persia, China, and throughout Europe, that whenever cholera visits a place in epidemic form, all the birds leave the locality, which seems, therefore, for the time being, under the curse of God. This is most remarkable in India, where certain birds constantly frequent the habitat of man, notably kites, vultures, crows, and the India *minah* bird. The vultures and crows are the natural scavengers of Indian towns and camps. Wherever soldiers are found on the line of march, wherever native fairs are held, these omnivorous birds will be found collected, as if by magic. But, let cholera only break out, and again, as if by magic, they vanish!

In 1846, at Karachi in Sinde, the 86th Regiment were stricken by cholera, and lost in ten days two hundred and fifty men. On the first two days of the visitation it was remarked that all the natural scavengers vanished, and it was also noted that when the disease disappeared they immediately returned. The case was equally remarkable a few years later, when a Highland regiment stationed at Hyderabad suffered to even a greater extent. My attention has again been called to the subject in the present year (1892), when Kullu was so terribly afflicted by this scourge. Mr. Donald, a large fruit-grower in the valley, suffered considerable loss from the depredations of the *minah* birds amongst his plums, apricots, and early peas, so much so that he placed some of the plum trees entirely at their disposal; that is to say, he took no measures to protect them. One morning early in May he noticed that the birds had entirely disappeared, and he was thankful that a certain amount of his fruit had been spared. But, alas! it was no matter for rejoicing. Within a few days he had nine cases of cholera on his estate, some of which proved fatal. The birds did not return till the epidemic had subsided. It is now considered safe to follow the birds, for where they are found congregated there will be no cholera.

PART-FIVE

AMONG THE LAMAS IN LAHOUL

CHAPTER XXII

LAHOUL is under British protection, and is bounded by Kullu and Chamba on the south, by Chamba and Zanskar on the west, by Ladakh on the north, and by Spiti on the east. It is triangular in shape, the river Chandra forming two sides of the triangle, and the Bagha the third. The inhabited portion of the valley is between ten and twelve thousand feet above sea level. The country is surrounded on every side by glaciers; the centre of the triangle is an enormous unexplored ice-field, and the cold is at all times intense. A glance at the map, however, will show its position; suffice it, therefore, to say that we crossed the Rohtung Pass (which I have already described) into Lahoul, on the 6th of May, 1892. At the top of the pass we took a drink of the water of the spring from which the river Beas takes its rise, hastily crossed the plateau at the summit, and commenced the descent. On the north side there was still a considerable amount of snow, which, however, was melting fast, and the slush we had to wade through was appalling.

We reached the rest-house at Koksur in due course, and, bad as it was, we were glad to get into it. Nature has provided shelters on this treacherous pass, in the shape of deep caves, which are much resorted to by travellers, and have doubtless often been the means of saving life by the protection which they give from the sudden and violent storms which are here so common. Some of these caves are large enough to contain even flocks of sheep and goats in addition to their masters, and such an one we found immediately in front of the rest-house, which we nicknamed the "hotel," for no sooner was it vacated by one drove than it was occupied by another. It was amusing to us to watch the different guests as they drove up their flocks of sheep, goats, and donkeys to the mouth of the cavern and unloaded, i.e. took off the pack-saddles, gave the animals a kick, and sent them off to seek what subsistence they could find. Then they covered the loads with a thin cloth as a protection against wet, and settled themselves down to feed as comfortably as if they were at the Grand in Northumberland Avenue. After all, life is a mere matter of habit.

In the autumn the summit of the Rohtung was carpeted with lovely wild flowers of every kind and colour, enjoying a short span of life, emerging from the snow in August to be but again covered by it in September. When we crossed, however, on May 7th, it was much too cold for flowers, or even any fresh grass, though there was comparatively little snow about. It was a dark, cloudy day, and we experienced no difficulty from the wind. The view at the summit was magnificent. In the Sonapani *nala* last year, an officer of the Scottish Borderers was lucky enough to kill three fine ibex in a couple of days. We were in a hurry to push on to Kailang, and the weather when we were there did not admit of our trying for them.

On the pass we met thousands of sheep in charge of the Lahouli traders, each loaded with its small saddle-bags containing rice, every grain of which has to be imported into the valley, as none will grow there. These sheep and goats make many trips to and fro during the time the pass is open. Exceedingly sure-footed, hardy, thriving on the short grass, brushwood, and whatever they can find, they are undoubtedly the best means of transport that could be devised. When halted they require no care, and the loss of a few now and again does not affect the trader like that of a mule or pony, which, moreover, is more likely to fall in the many difficult and dangerous places which have to be traversed. It was a curious sight every now and then amongst the flocks to see a ewe laden, not with grain, but with a small lamb in each saddle-bag! Poor little things, what a life to look forward to! True that they are not killed for food, but are worked till they drop. Which is better, the long life of toil of the Lahouli sheep, or the short and merry one of his brother in the green pastures of England?

We waited in the rest-house at Koksur for our tents and baggage, which took hours longer to cross the pass than we did. This little rest-house, 10,338 feet above the sea, exposed to the notoriously cold and searching wind, was eight years ago the scene of a terrible catastrophe, when one October day four hundred sheep perished in and around its verandah. The *guddis*, or shepherds, were overtaken by a snowstorm between the rest-house and a place called "Purana Koksur," four miles up the valley. The *guddis* fled for their lives, leaving their herds to their fate. Two thousand perished in the night, remaining buried under the snow till the following spring, when it melted. One shepherd alone stuck to his flock, and brought four hundred to Koksur with him, but as there was no shelter for them there, they likewise succumbed. Since this year of misfortune the *guddis* always quit Lahoul in the middle of September, to avoid the risk of another such storm.

As there were no more rest-houses between Koksur and Kailang, we ordered twenty coolies to come in the morning and take our tents and some kit on to the next camp, so as to make sure of some shelter on arrival, as it was a long march. D. was walking up and down after breakfast, when he suddenly drew my attention to some half-dozen young ladies who were making a rush for our kit in the verandah. He remonstrated, but was given to understand that they were the coolies who had come to carry our loads. Some of them could not have been more than fourteen or fifteen years old, and I suggested that it was impossible for them to lift the loads, much less carry them fourteen miles, especially as we always arrange that each load is not less, but usually considerably over 60 lbs. The young ladies quickly undeceived me, however. As soon as the loads which were to go had been pointed out, they tied their ropes round them, sat down, settled them on their backs, rose with a little assistance, and then, laughing and cheery as if they were going to a wedding, strode off to carry them fourteen miles for the magnificent sum of fourpence. Moreover, instead of selecting the gentle zigzag which had led from the rest-house to the main path above, they climbed straight up the mountain side, without an apparent thought of the loads on their backs. The few men among the gang were far more particular. Selecting the lightest loads for themselves, and leaving the heavy ones for the women, they grumbled even over the former. A month later we were told that no men at all would have been procurable, as they all would have quitted the country for Ladakh, to return with borax and other merchandise.

These weight-carrying females merit a special word of description. Of the pure Mongolian type, they were short, thick-set, though well made, with small hands and feet, and a complexion by no means dark, though leathery from constant exposure to hot sun and icy winds. Attired in a frock of grey homespun, with a piece of red let in down the side, and girdled with a rope of goat's hair, they wore very tight home-spun pyjamas. From one side of their waist hung a long chain of brass and imitation turquoise beads, ending with a brass spoon, which they always made use of at feeding time. Their hair was plaited in very fine strands from the forehead to the back, where it was tied with goat's-hair thread and fastened with a silver ornament. On the top of their head they invariably wear a kind of silver saucer. They were also adorned with necklets of bits of jade, coral, coloured bone, inferior turquoises and beads of sorts, and they were dirty beyond description.

The men were even more hideous in appearance than the women. Dressed in a thick, warm woollen coat or sheepskin, with the wool inside, they wore pyjamas of felt fastened round the leg like a Kashmir *putti*, with a garter winding upward from the ankle. A sheepskin cap, likewise with the wool inside, and with a curtain behind to protect the neck. Felt boots with soles of sheep or goat-skin, completed their costume. The garments are never washed, and the natives themselves are said to wash but once a year, on the day of their holiest festival.

We had been given to understand that one of the advantages of Lahoul was the absence of rain there. During the three days, however, which we spent at Koksur, it hardly ceased raining or snowing. Consequently we had no chance of exploring the glaciers of the Sonapani *nala*, as we had set our hearts on doing. It was intensely wretched at Koksur. The chimney of the rest-house smoked, so that we could not light a fire, though the thermometer was barely above freezing point. Moreover the only wood to be got—and the supply cannot last many years longer—was a diminutive silver birch, which declined to burn without incessant blowing, emitted more smoke than any wood I have ever used, and left no ashes, merely dust. Further up, we only could get the pencil cedar (*Juniper excelsa*), which they said possessed the disagreeable peculiarity of making everything that is cooked with it taste, but we did not find it so.

While at Koksur we took the opportunity of examining the famous grass of Lahoul, and I fancy we have discovered the secret of its fattening properties. It is largely composed of wild carrots. The effect of these on horses and sheep is simply marvellous, and so greedy are they for it, that when once let loose, it is with great difficulty that they can be collected and caught. When the Lahouli traders want to load up for the march, they lay lines of rope at right angles to the heaped-up loads, each line secured at either end by a peg in the ground. The sheep and goats, each of which has a rope collar, are driven towards the lines and kept in a flock till caught one by one and fastened by the collar to the lines, one on each side. When they are all secured, the loads are put on their backs and they lie down till ready to start.

Detained at Koksur by the continuous sleet and snow, we dared not move on, trusting only at the next camp to our tents. A snowstorm might have broken our ridge poles, leaving us under the canopy of heaven in this inhospitable climate. To make matters worse, we were running short of money, and knew that some days must elapse after our arrival at Kailang ere we could get some more.

A batch of visitors arriving from Spiti, with horses they were taking down country for sale, enlivened us for a while. Though no larger than donkeys, I have called them horses, they were so exquisitely shaped; and I tried to do a deal, but did not succeed. The men, a hideous crew, just like Chinamen, but cheery and sociable to a degree, settled themselves, to our horror, in the verandah of the rest-house. This we could not permit, feeling convinced that their long woollen coats were simply "crawling." It was hard lines on them, however, to have to put up with the mere shelter of an exposed rock, the cave "hotel" being fully occupied.

It is simply a disgrace to the British Raj, that, at a place like Koksur, 10,338 feet above the sea, the necessary halting-place of all travellers ascending or descending the Rohtung, that there is no *serai*. These are found in plenty down country, where sleeping in the open is no hardship, and abound in every native state in India. Here, some protection from the blinding snow, the fierce gales, and the intense cold, is necessary for saving life, and in consequence, when pressed, the travellers break into the rest-house provided for the *sahibs*, lighting fires on the floor, and doing much damage. I think myself that they are quite justified in doing so. But stone is plentiful, sufficient timber for roofing lies to hand in the old beams which have been taken out of the bridge, and the *serai* ought to be begun at once.

Leaving Koksur at eight a.m. on May 10th, we reached Sisu in five hours. The views along the road are superb, glacier after glacier and *nala* after *nala* coming into view along the left bank of the river. A great take-off, however, from the thorough enjoyment of travel in these parts, are the sickening cases of cruelty to animals that one constantly meets with. We passed flocks of sheep laden with grain, and saw many sheep barely able to creep along, and many left to die on the roadside. The poor beasts of all kinds appear to succumb to pneumonia, brought on by exposure and insufficient food. We camped at Sisu, on a small plateau shaded by a few pollard willows, which would have been pleasant had it been further off the village. Hardly had we got our tents up when down came the rain, and we were thankful it was not snow. It rained all night, but we experienced no inconvenience beyond that of keeping alight a cooking fire. As wood was very dear—two rupees for a hundred and sixty pounds weight—we did not indulge in the luxury of a camp-fire. As usual, milk was not easy to procure. The milk-givers of Lahoul are a cross between the Yāk of Thibet and the small hill cow, and queer-looking creatures they are, with big horns and long bushy tails; but they give delicious milk. The difficulty, however, is that there are not enough of them. The true pure Yāk, though occasionally met with lower down in winter, will not live at a less altitude than fourteen thousand feet. The hybrid appears well suited to the climate of Lahoul, being able to stand the heat of summer and the great cold of winter, which latter invariably kills the cows from down country.

We had crossed the Chandra at Koksur by a *sungha* bridge, which had cost over nine thousand rupees to erect. The former bridge, five miles lower down the river, was at a more suitable spot, and saved time in the long march to Sisu. But it had been carried away, and an iron one, costing twenty-five thousand rupees, which was being built to replace it, collapsed, owing to inferior workmanship, just as the coolies were laying the platform, and eleven of them were drowned. The present *sungha* bridge was not exactly a nice one to negotiate, for it was very high out of the water, very narrow, and boasted no handrails. Each year in October the planking which forms the platform is removed, that it may not be broken down by the weight of the winter's snow, and the river is impassable till after the first heavy fall, when there is no difficulty in finding snow-bridges.

This stream rises at the foot of the Bara Lacha Pass, flowing east, till at the Shigri glacier it makes a sudden bend to westwards, a turbulent, muddy stream like pea-soup. After being joined by the Bagha river, which flows down the Kailang *nala*, it assumes the name of Chandra-Bhaga till it reaches Kishtwar, when, under the name of the Chenab, it is known as one of the five rivers of the Punjab. Twenty-four years ago one of the glaciers below Shigri gave way, and the river was dammed with debris for seven days. In those days, even in the summer months, there was no postal service in these parts, so the news could not be sent down country, and when the dam burst it caused disastrous floods in the low country, and much loss of life.

We started early on a fine morning for our double march, reaching Gundla at eleven. After waiting an hour and a half for our coolies, who were to be changed here, we went on, leaving word that if they wished to be paid they must send one of their number on to Kailang for the money. Between Sisu and Gundla we saw across the river the tent of a *sahib*. This we afterwards ascertained belonged to Mr. P—, of the R.A., who had commenced shooting in the Seetee *nala*, the best in Lahoul, and was trying all the *nalas* on that side of the river up to Tandi, opposite the Kailang *nala*. Up to the time we heard of him he had secured a couple of brown bears, but had had no luck with the ibex. The fame of his commissariat arrangements, however, was brought to us.

He had contracted with his servant to give him three meals daily, including soup twice, for the sum of eightpence per diem! This arrangement had been going on for some weeks, and he expressed himself perfectly satisfied. D. humbly suggested that he at once be made member of the committee for the better feeding, without increase of expenditure, of Thomas Atkins.

Between Sisu and Kailang we passed several long lines—some as much as twenty yards in length—of stones. At each end of the lines is a pile of stones in the shape of a Thibetan wooden prayer-wheel. On each of the stones is cut the Thibetan prayer or *mani-door*: "Om mani padmé hum!" ("O God, the flower in the lotus!") These piles go by the name of *mani panés*, and are erected by some religious native as a thank-offering for a great mercy vouchsafed him, such as the recovery from a serious illness, the birth of a son, or a good harvest, or are added to as votive offering for some special object. These stones with the inscriptions are bought from the *lamas*, and as there are thousands of them in each line, each inscribed with the same words, the *mani panés* must be costly undertakings.

I took a fancy for some of these stones, but fearing that it might be considered sacrilege, I asked the opinion of the Lahouli who was carrying the lunch-basket before taking one. Reassured by his reply: "Of course, take as many as you like. Who cares?" I carried off two. It is a curious thing that when Buddhists pass these *manis* they always keep them on the right.

We reached Kailang about five o'clock.

CHAPTER XXIII

THE village, or I suppose I should say the capital, of Lahoul, is situated on the right of the Bagha river, about a thousand feet above it, and commands an extensive view of glaciers in all directions. The country immediately around it is well planted with pollard willows, which thrive in these regions, and are much appreciated both as firewood and as forage for cattle, when nothing else is available. The trees are planted in groups of four, tied together with grass ropes and bound round with bits of old clothes to protect them from passing flocks, which greedily devour the bark. This way of planting protects them against the fierce winds that prevail, and in a few years they grow together into one substantial tree. Every four years the upper branches are lopped for fodder.

I had succeeded in obtaining fifteen gold *mohurs* of various dates and reigns and of considerable value. Being now on the high-road between Central Asia and India, we had hopes of picking up some of the gold coins of Bokhara called "tilas." They are coined by the Amirs there, and weigh about half a *tola*, and are much sought after by the natives, as all gold coins are, for neck ornaments. At Dalhousie, a few years back, we managed to buy some gold Dutch ducats, about the same weight as the Bokhara *tila*, but easily distinguishable from them by the figure in armour on one side and the Latin inscription on the other. There are said to be thousands of them in circulation in India, but I have very seldom come across them. When in Kashmir I bought some Kashmir "chilka" rupees for buttons. These bear the Christian monogram "I.H.S.," the origin of which has recently been explained to me by a well-known Indian coin collector, as follows: The Maharajah Goolab Singh tried all sorts of signs on his coins as talismans. A renegade native Christian who had entered his service was asked what was the most powerful charm used by the Christians. He immediately bethought him of the letters which meet the eye on first entering a church, embroidered on the altar-cloth. These the maharajah ordered to be struck on his coins, and there they remain till this day, though not a soul in Kashmir knows their meaning!

We were much inconvenienced on reaching Kailang by the non-arrival of our beds. Unless we took the precaution of putting the largest loads on the strongest coolies, and of seeing them actually off, they were invariably the last to turn up. On this occasion everything turned up except the all-important bedding, a uniform case of mine, and a portmanteau of D.'s, which always formed one coolie load. When they did arrive they appeared borne by two children of not more than nine or ten years of age, and how they turned up at all must forever remain a mystery. There is no society for the prevention of cruelty to children in Lahoul. Considering the superior physique of the women, I have no objection to their carrying loads. But there should be a limit as to age. One little women, one of the first of the gang to come in, we found on measurement to be but four feet six inches. Her load was seventy-eight pounds, and she simply laughed at it.

We busied ourselves all the next morning in drying our soaked tents, in levelling the ground, and in pitching them, preferring our canvas home to the rest-house with its broken windows and doors, and its smoky chimneys which prohibited a fire, especially as, after the first three days, we had to pay four rupees a day or one hundred and twenty rupees a month for the privilege of occupying one room! In the afternoon we called at the mission-house. It is a Moravian mission. Mr. Heyde, the missionary, has been in Lahoul for thirty-seven years. The house is pleasantly situated as regards the view, but unfortunately is built exactly under the village of Kailang, which is as unsanitary as all native villages, and the bungalow is, to my mind, undoubtedly unhealthy. The poor Heydes have lost many of their children of diphtheria and typhoid. Both are charming people, kindly and homely, and we enjoyed some of the most delicious bread and butter I have ever tasted out of England. We spent a couple of hours with them discussing Lahoul and

Ladakh, the country beyond, with which Mr. Heyde is well acquainted. An excellent Thibetan scholar, he holds services on Sunday in that language.

The next day we received a visit from the Thakur of Lahoul. He is an unimpressive little man, but wields immense power in his own country. The natives care nothing for Europeans, but they quake in their shoes at the approach of this little tyrant. As the owner of large numbers of ponies, some hundreds, he would be invaluable to Government should troops at any time be sent through Lahoul to Ladakh. The secret of his large stable is that every colt born in the country goes to the Thakur. He presents the owner with a new cap, and pays for the keep of the mare for six months in exchange. Fillies remain the property of the owner of the mares.

Our Spiti friends, with their ponies, having turned up again, I succeeded in doing a deal, and became the possessor of a likely looking little mare, on which I should make money should I succeed in getting her down country. The price I paid was twenty-six rupees and the "old hair trunk" (as I called the Ladakhi quadruped D. had bought in Kullu), and a few cheroots thrown in. The mare we have christened "Polly," being the short for "polyandry," which is the custom of the country. We gave her a feed of corn, but she did not know what it was till we put some in her mouth, when she devoured it greedily.

Though my sympathy had hitherto never gone out to coolies, considering the loads they carried and the value of money in the country, nevertheless I pitied the people called upon to perform *begar* round about Kailang. Thence they are taken to carry food and wood and fodder for travellers as far as Rukchen, ten marches off, or a hundred and fifty miles on the road to Ladakh, whence, of course, they have to return, making a distance of three hundred miles for three shillings and fourpence. We should hear a good deal about slave-driving were we to ask an English man to walk a tenth of that distance for the same money. Even more unfortunate are the coolies who live in the neighbourhood of some high pass, over which they have to carry the chattels of travellers. The marches are long, and the labour of ascent and descent when carrying a load is very great. Further, they suffer much from snow-blindness. While we were at Kailang an Englishman came over the Kookti Pass from

Chumba, and had to leave all his coolies on this side till they had sufficiently recovered from their blindness to return to their homes.

The same thing has often happened to us in Kashmir, inspite of every precaution—such as crossing the passes in the night-time, and of protecting their eyes with green sprigs of fir. A drop of tincture of opium in the eye will temporarily relieve the pain, but it takes some days to effect a cure in the sight, which is sometimes permanently injured. In the absence of goggles, it will be found an admirable plan to paint thickly with charcoal beneath the eyes and the bridge of the nose. If this is done, snow-blindness will to a certainty be avoided. I recommend it strongly for the protection of the eyesight of servants and coolies who are not provided with goggles.

Ploughing and sowing were in full swing around Kailang when we arrived. The female population, which outnumbers the male by a hundred and seven to a hundred, were hard at work—most of their lords and masters being away trading—and were ploughing with a yoke of hybrids, keeping up an incessant chant of "Yo! yo!" much like sailors hauling at a rope. The snow had only vanished with the end of April, and in these high regions they get but one crop a year—in September—chiefly buckwheat and barley, and these only with careful irrigation, owing to the precarious rainfall. Though the population is but 2.5 to a square mile, so little land is available for cultivation—nothing will grow over twelve thousand feet—that the inhabitants depend much on Kullu for food. Within the last forty years the Moravian missionaries have taught them to grow potatoes, and very fine crops they are, hard to beat in England; but to make up for lack of other vegetables, they eat weeds greedily, chiefly dandelion leaves and those of the stinging nettle, cooked like spinach. The only fruit in the upper valley is the wild strawberry. The poor people drink *chang*, made of an infusion of water on boiled barley; the richer, tea, cooked with clarified butter and salt, and concocted with a churn kept in every house for the purpose. They also distil a kind of muddy whisky, D. found not unlike "potheen," and very inferior to the whisky made by the Ladakhis. The poor live on ground buckwheat, eaten as we eat porridge, or ground into flour and made into bread by being mixed with *chang*. The well-to-do people eat rice boiled with ghee, and mixed with sugar and dried fruits. It is against the creed of the Lahoulis to eat the flesh of cattle, but some are killed on the sly, and also many sheep, which are dried, and keep, at all events through the winter, in this climate. They appear to have no scruples, however, against devouring animals that have died a natural death; and the low-caste Lahoulis will eat even beef, and bear's or panther's flesh.

Fodder is plentiful. By the sides of the irrigation water-courses the grass grows high in the summer, and makes excellent hay. The barley straw is collected for winter fodder, except what is required for making grass-shoes, and in the spring they use the leaves of the pollard willow. So improvident and lazy are the Lahoulis, however, that in a late spring there is never any margin of forage in store, and many poor beasts die of starvation. Like all the other natives of Hindostan, it seems an article of their creed only to give them enough to keep body and soul together, and so they are obliged in self-interest to feed them up a little before the ploughing begins.

The Lahouli villages we saw in the upper valley looked at a distance like one huge house. Four or five substantially built dwellings are grouped together, with inter-communication, under one roof of thick slabs of slate, both for economy of space, and also for convenience during the winter, when the country lies under ten to twenty feet of snow. On these flat-paved roofs, made strong to withstand the weight of snow, is stocked the store of hay, and inside is piled the wood. During the winter months the Lahoulis seldom leave their villages, but spend the day in sleeping, drinking *chang*, gambling, and dancing. They are born gamblers. Nearly every Lahouli carries dice about him. Their dancing is prettier than any we had yet seen in India, Men and women danced together, something not unlike a country-dance at home, and very superior to the ordinary terpsichorean "divarsh'on" of the hill-men, to the sound of lutes and tom-toms. Though they do not play the kind of aboriginal polo I have seen in Kashmir, far away towards Gilgit, we were told they go in for horse-racing.

Polyandry is the rule and not the exception in Lahoul. It is the custom for several brothers to live together with one wife, which has the advantage of preventing the subdivision of estates, and also of keeping down a population already dependent on outside help for support; though the moral effect on the people is disastrous. As a consequence of polyandry, a certain number of girls in each family are made nuns. The nun of Lahoul is, however, a very different creature to her cloistered sister in Europe. Wearing short hair, and dressing like men, they are only distinguishable from boys by the brick-red sort of Balaclava cap they wear. At liberty through the summer, and living in their own homes, in the winter they retire to the monasteries, where they are educated in a very elementary way, learning the Thibetan alphabet, weaving blankets, and making themselves generally useful. The *lamas* in the monasteries are of the Drukhpa sect, and though all are married, are a degraded people. All other sects of Buddhists are bound to celibacy.

At Kailang we fell in with Mr. M—, on his way to the Tso Morari after *ovis ammon*, and on the 21st of May we started to explore the Bailing *nala*, just behind Kailang, in order to choose a camping-ground, should we decide to remain in these parts. The *nala* was so steep at its entrance, that we had to make a detour of half a mile to the east, and ascend by a very fair zigzag road leading to the Lama Serai—pronounced "lamassery," and called by the Buddhists *Gonpa*, which means, in the Thibetan language, "a solitary place." In such their monasteries are always built. The Heydes had made the road, as their farm lands completely surround the monastery, the lamas of which appear to be content with a house to live in, and not to go in for cultivation. The Lama Serai was not unlike an ordinary villager's house, except that it was two-storied and covered with flags, which had once been white, and it stood some twelve hundred and fifty feet above the village.

A hundred feet beyond the monastery we came upon an aqueduct or *kool*, by the side of which was a path some eighteen inches wide leading into and up the *nala*. The *kool* was cut in the mountain side with some amount of engineering skill, and upon inquiry proved to be also the work of the *Padre Sahib*, i.e. Mr. Heyde. The answer is always the same. A good road, a good watercourse, a good house, a tidy field, fat cattle, a clean native—the abnormal product is always due to the beneficent missionary. Unsuccessful he may be in his conversions to Christianity, but his efforts and his example are not thrown away on this poor people.

After following the water-course a couple of miles, passing through a so-called forest of pencil cedar—though there were not more than half a dozen trees to the acre—we came upon the first *thach*, or feeding- place, in the *nala*. It was a veritable Stonehenge. But if the patches of grass were few and far between, they made up in quality for what they lacked in quantity. We were now at an altitude of thirteen thousand feet, and it was snowing hard, no unusual occurrence, for we remarked that on this, and every *thach* we passed higher up, the shepherds had erected stone shelters with wooden roofs. As the beams must have been brought from great distances, no Lahouli would have given himself the trouble to do so unless it were absolutely necessary for himself. For the sake of his flock, never! On the Kailang Gong *thach*, about two o'clock, we went into one of these shelters, as it was still snowing, to eat our biscuits and drink our whisky. There were no sheep in the *nala* yet, the wind off the glaciers by which it is surrounded being still too keen for them. It cleared a little after lunch, and further on up the *nala* we got a full view of the Bailing glacier at its head, and bringing out our glasses, took a look round. Before long we spied a herd of four ibex feeding on a slope some three thousand feet above us, and on the opposite side of the ravine. There were two males, but the heads, as far as D. could make out with his telescope, were not more than twenty to twenty-four inches. He calculated by taking the length of the head from the ears to the tip of the beard and comparing the horns with this length. As it was now past three o'clock, and would have taken two hours to get up to them, we decided, much against my wishes, for I thought the heads larger, to leave them alone. A valueless trophy was not worth the trouble of a two hours' descent to camp in the dark over a most unreliable path.

Great difficulty of breathing, the result of the rarified air, was remarkable here in all of us—myself, D., and the man we had brought with us from Kullu. At greater elevations we had not been so inconvenienced. I account for it by the fact that we were quite cut off from any winds coming from the plains, and entirely surrounded by glaciers. We could not walk even a few yards up a slight incline without stopping for breath.

Far up the *nala*, at the height of fourteen thousand feet, we found large plots of dwarf rhododendron with a pretty yellow flower, a hardy plant, growing up to even fifteen thousand feet. Lahoul is especially rich in rare flora, but contains no ferns or mosses, probably owing to the exceeding dryness of the air. At Kailang we found a pencil cedar (*Juniper excelsa*) thirty-three feet in girth, and which I should imagine is one of the largest specimens in the country. Mr. Heyde, the missionary, has kept the string with which it was measured.

CHAPTER XXIV

THE religion of the Lahoulis—if they have any—is Hindoo Buddhism, largely permeated with demon-worship. Originally demon-worshippers, Buddhism obtained a hold of them about A.D. 600, since when it has been the recognized religion. But constant intercourse with Hindoos for trading purposes, and the immigration of several into the valley, has produced the Hindoo-Buddhist religion which we now find. The houses of the worshippers of *Cha*, a demon, are distinguished by a round pile of ibex horns, like a chimney stack, on the roofs. These are collected after a severe winter in considerable numbers when the snow melts, showing how many of these beautiful beasts annually perish in the avalanches. Some of the horns are very fine, one I measured with the tape to be forty-five inches, and there were doubtless several larger. Each mountain peak is dedicated to some saint, whose name it bears. Immediately opposite our camp at Kailang was a serrated range of seven peaks, named after the seven principal disciples of Buddha. Further down, was a peak eighteen thousand feet, on which one of the saints is said, traditionally, to have sat and fasted for forty days and forty nights. But the extreme cold, added to the long abstinence, must have overtaxed, I should imagine, the constitution of even a Lahouli saint! Above our camp was a cave where is deposited an idol who is supposed to be responsible for the giving or withholding of rain. When his offices are required a *lama* goes in and intercedes with him. Should the deity prove unpropitious, he is turned upside down, pelted with old shoes, and abused freely.

Nevertheless, there is a good deal of outside religious show in Lahoul. Witness the *manis* before described, the ubiquitous monasteries or *gonpas*, and the *chortens* or minaret-shaped stone buildings, erected in so many of the fields in memory of some saint or *lama*. In addition to demon-worship, the Lahouli mind is deeply tinged with belief in witches, sorcerers, and in the evil eye.

Mr. Heyde was of opinion that there were three distinct kinds of religion in Lahoul—Buddhists, half-Buddhists, and Hindoos, and that they each and all stick to the various faiths with a tenacity which gives little hope for the spread of Christianity among them. It must be cruel work for him to see such small result for his years of labour. The schools which he started in the country have now all been closed but one, and for the following reasons. No native can understand a good action being performed without a *quid pro quo*, even when the benefactor is Government itself. The schools started by the mission originally received the support of Government, and the Lahoulis, to curry favour with the authorities, sent their sons to them. After a while, however, perceiving that they reaped no immediate benefit, and imagining that Government was taking some advantage of them, they gradually withdrew the children and combined to petition Government to excuse them *begar* (forced labour) if they sent their sons to school. The petition being, of course, refused, they have since so pestered the officials for that, or some other remuneration, that in December last the schools were closed by order, except the mission school, which is only largely attended during the winter months.

On Sunday afternoon we went to divine service at the mission house. The morning service was more in the shape of a discourse, and less easy to follow in Thibetan than the afternoon one, which is on the lines of an English service, though also in Thibetan. The room set apart for the chapel was plainly furnished with forms, a reading-desk, and a harmonium; lighted with twenty wax candles, and warmed with a stove—no mere luxury in a place where the thermometer often falls in winter to 5° or 6° below zero (Fahrenheit). The congregation consisted of eighteen persons—men, women, and children, and there were absentees through sickness and other causes. All sang, and sang well, and the service, which lasted about an hour, commenced and finished with a hymn, and was interspersed with others. The native Christians wore their national costume, the girls the usual *pèrâk* or head-dress of rough turquoise, which became them well. We were both surprised at their good looks, such regular features, large eyes, well-marked eyebrows, good teeth, and especially at their unusually fair skins. At first we were at a loss to understand the latter, till it transpired in the course of conversation at tea over Mrs. Heyde's delicious bread and butter (the first and best we had tasted for nearly two years) that she insisted upon the frequent use of cold water. Soap, indeed, is absolutely unknown in the country. When we reached Kailang and sent our clothes to be washed, we were asked for soap to wash them. As we could spare none, we had to borrow from the mission, and send seven marches down to Sultanpore, in Kullu, for some to repay it.

I asked Mrs. Heyde at tea about the peculiar head-dress I had noticed the women wearing in church. It appears that only married Ladakhi and Thibetan women are supposed to wear the *pèrâk* head-dress. But I have seen it donned by females of all ages, so suppose that they are not particular on this point. But unmarried women wear only one or two turquoise beads in their hair.

The *pèrâk* consists of a leathern, or more often, a scarlet cloth band, about one and a half inches broad at the forehead, six inches broad where it covers the centre of the head, and three inches broad below the nape of the neck. Why does a miller wear a white hat? The answer to this fine old crusted riddle will probably equally explain the origin and custom of the *pèrâk*. Antiquarians, however, see in it a survival of the old days of universal serpent-worship, for the *pèrâk* somewhat resembles in its shape a well-fed snake. The idea is a happy one, and is open to anyone who likes to accept it as the solution of the riddle; but I can hardly accept it myself.

The *pèrâk* is studded at intervals of one or two inches, with very coarse turquoises, of colours varying from deep green to the purest blue. The people themselves describe them as the old and the new turquoises. We should be inclined to class them, perhaps wrongly, as imitation and real. The latter are exceedingly rare and can be recognized at once by their pure colour. What they call the new turquoises fade, turning to a greenish hue after exposure to the light of day, or a prolonged contact with the flesh of the head and neck. But the old preserve their colour under any circumstances. It is improbable, however, that any of them come from the world renowned mines of Nisharpur or Meshed, in Persia, the finest stones from which are annually sold at the fair of Nijni-Novgorod and at Moscow. The Thibetan and Chinese turquoises, though numerous, are of but little value, and do not retain their colour. Not that this seems of much importance to the wearers of the *pèrâk*. In the same way the natives of the outer Himalayas, who prize coral exceedingly, are quite content with the common deep red kind, which we do not appreciate at all. Neither do the Thibetans care for stones without a flaw. They look upon these with suspicions as to their genuineness, and readily purchase the inferior coloured ones, even if liberally covered with the black spots which we consider utterly detrimental to a valuable stone.

The largest turquoise is always worn on the extreme end of the *pèrâk*, on the forehead, and is sometimes as large as, and not unlike, a thrush's egg. I often tried to buy one of these very large stones when of a good colour, thinking that a good stone might be cut out of it. But the Thibetan ladies declined to part, probably because, being uncertain of their value, I did not offer them enough. After all, turquoises are not so easily found in these parts as some people imagine. A sportsman passing through here recently, on his way to Zanskar, asked Mr. Heyde, in all gravity, of a place where he could pick them up, literally, on or near the road.

I have before said how difficult it was to procure a fowl in Lahoul. They were never seen unless brought into the country by the Heydes from Ladakh (between two hundred and three hundred miles), and ducks were practically unknown. Now, before marching into Lahoul, we had been lucky enough to get four ducks from a friend who had come from Simla. We carried them along with us, and at Kailang they created quite a sensation; even Mrs. Heyde asked permission to bring up the children of the mission school to see them. I merely mention this to show what a primitive people we were amongst.

During the summer months only, there is a post twice a week from the outer world, and while we were at Kailang we heard that the rajah of Kullu and three of his wives had died of smallpox. He would never consent to be vaccinated, and his evil example influenced his people, who, by every means in their power, evaded the law of compulsory vaccination, which has done so much to keep under what was formerly the greatest scourge of this beautiful country. People in England can have no conception of the rapidity with which an epidemic spreads in these southern latitudes, and of how utterly negligent the people are of the most ordinary precautions. The very clothes in which smallpox or cholera patients have died are worn by the survivors without even being washed. The bodies of the poor are often only half burned and then thrown into the nearest river, which is also used for drinking purposes. When possible, the corpse is really burnt, the ashes thrown into the river, or collected and made into a cake with flour and deposited in a chorten in the fields. In Spiti and Ladakh the bodies are not unusually exposed to be eaten by wild beasts, or cut to pieces and given to the dogs. Happy is the man esteemed whose remains are devoured by the sacred dogs kept for the purpose by the *lamas*. A reincarnation in a higher form will be their reward. I think it is Monier Williams, in his lectures on Buddhism, who calls the dogs "the mausoleums of Thibet!"

The mail which came in on the Queen's birthday (God bless her!) brought us confirmation of the death of the rajah of Kullu and two of his wives. The third met with an even more tragical fate. The rajah had recently purchased an idol in Calcutta, which he brought back to Kullu and set up in his castle for his women-folk to worship. When he lay dangerously ill one of his wives prostrated herself before it and prayed that he might be spared. As she lay praying, word was brought her that he was dead, whereupon she beat the idol mercilessly with her shoe, and taking a sword which was suspended over him, cut her own throat with it. She and the rajah were cremated together. Matters in Kullu were indeed in a bad way. Not only were cholera and smallpox rife, but a severe drought had resulted in the withering of the crops and the starving of the animals. The infected bodies were thrown into the river and the diseases spread. Some friends wrote to us that for days they had not been able to get a coolie to work for them, as a story had been invented that four of their servants had died, and that they had caused them to be buried in the forest. As a matter of fact none of their servants had died.

Captain M—, who had passed through Kailang a week before on his way to Leh, returned suddenly, having been unable to get over the Bara Lacha Pass, on account of a snow storm rendering it impassable for his mules. I have heard of mules being got over soft fresh snow by throwing down blankets for them to walk on, but apparently this dodge was not tried. A Mr. M—, however, who left later than the captain, did not return. Perhaps he left his mules and took only his coolies over. Young Mr. P—, of the eight *anna* a day fare, now passed through, and we had the pleasure of inquiring about his commissariat. He found his fare rather monotonous, however, for he had sent back to his regiment in Central India for thirty-six pounds of jam to be forwarded to him immediately. He passed on to Chang Chenmo, ten marches beyond Leh, and our best wishes went with him.

We found ourselves rather in a fix at Kailang for lack of money. D. had sent back to Sultanpore, where he kept a deposit, for some, only to learn that his banker, Muttra Pershad, had closed his shop and fled to the mountains to escape the cholera. The Thakur of Lahoul, however, kindly obliged him with fifty rupees, on his simple note of hand, and asked for no receipt. Natives are very confiding. I remember I was once at a Delhi jeweller's buying a sapphire. As it was getting too dark to see, I hastily put a handful of stones in my pocket and walked off, saying I would return the next day. The jeweller made no objection, and though it was a few days before I returned, he made no inquiries about his valuable stones. I asked him if he had ever been taken in, and he replied, "Never!" Well for him that Delhi is remote from some of our European capitals!

CHAPTER XXV

ON the 25th of May we paid a visit to the Kailang Gonpa, Standing on a spur of the mountain, nearly two thousand feet above the village, it commands a superb view of the entire Bhaga valley, with the glaciers at its head, and also the glaciers on the left bank of the Chandra. The air was crisp and invigorating to a degree, and any sickness must be the result of the inhabitants' own filthy habits. Just before reaching the edifice we passed a *stupa* recently erected, as the still clean whitewash showed. We were told it contained the ashes of a very holy *lama*, who departed this life only last year. In what the holiness of a *lama* consists I fail to understand. They are of the world worldly, ignorant, immoral, and drunken. Their only good trait is tolerance, as inculcated by Buddha, who says, "Never think or say that your own religion is the best, and never denounce the religion of others."

At the entrance to the monastery we were met by a *lama*, who had evidently been left as care taker, while the others were away in the fields working during the summer months for the small wage which is to support them during the winter. Except the neck-chain of a hundred and eight beads, there is nothing to distinguish a *lama* from any other Lahouli. We commenced our inspection at the southern end of the lower story which contains the temple. On the way we passed a small grass plot, with a may-pole in the centre, which was shown to us as the scene of the annual *lama* dance. Our *lama* guide pointed out to us the chief *lama's* seat, a stone slab, irreverently throwing a stone to indicate it.

The temple is a room within a room. Round the inner room or temple, runs, on all four sides, a three-foot passage, where the *lamas* perambulate, meditating, and turning the hundreds of prayer-wheels which hang on the inner side, muttering as they walk: "Om mani padmé hum." The wheels are highly coloured, and the cylinders contain prayers in Thibetan character. The *lamas* always pass the wheels on the right; and woe to him who accidentally turns one the wrong way. The walls of the outer room are gorgeous with florid paintings of saints and demons, and evil beasts, after the manner of Hindoo temples, showing the semi-Hindooism of the Buddhism of the country. Buddha, in fact, having commenced his teachings in countries where Brahminism had already secured a footing, and, finding that he could make no

Gonpa or Lama Monastery at Kailang

headway, was compelled to make a compromise, the result of which is still remarkable. Inside the temple, on a raised dais, stood a statue of Buddha, and those of his seven principal disciples. In front of him, eternally burn two small lamps, the only illumination in the temple, so that it was difficult in the "dim, religious light" to make out exactly what it contained. We discovered, however, paintings on silk, yak's tails, drums, masks, swords, and cymbals, all which properties were in request on great occasions, such as the annual religious dances to which I have referred, and which are called *cham*. Among the paintings was a striped creature, which might have been anything, from a pig to a zebra, and which our guide pointed out with great pride as "a tiger of your country," and seemed much gratified when I assured him that it was a marvel of the painter's art. It is quaint that all these hill-people have an idea that England is somewhere in the plains of India, and that, with the exception of a few *sahibs*, all the English people are black. They cannot understand white men doing menial work.

Above the temple was the abode of the chief *lama*, profusely adorned with paintings, and with a gilt and plaster representation of Buddha at one end. It communicated with the temple below by a shaft, and all round the room were a hundred and eight pigeon-holes, each containing a religious book. It was, therefore, the monastery library. These books were purchased from the monastery at Lhassa about three years ago for fifteen hundred rupees, subscribed by the *zemindars* of Lahoul. The hundred and eight volumes are all one work, and as Buddha himself left no writings, must all have originated from some learned Brahmin. The matter of the volumes is of very little importance to the *lamas*, who read, but do not understand them, sitting in a circle, each taking a leaf and reading simultaneously, and evidently neither knowing nor caring what they are reading about. The books themselves are mere bundles of loose sheets, printed from engraved slabs and laid one over the other, folded in a piece of linen, with a strip of silk at one end, on which is embroidered the name and number of the work. Above and below are boards to keep the sheets straight, which are bound tightly together with a leather thong.

Outside the chief *lama's* room was a row of cells for the inferior *lamas* and novices. But they were all away working in the fields, and the cells locked up. While we were going round, another *lama* appeared, accompanied by "a young thing." Whether it was a boy or a girl we could not decide. Upon inquiry it turned out to be a girl—a nun—left behind as cook to the two *lamas*. There were two houses behind the monastery where the nuns live in the winter, but we had no time to go over them. The monasteries are supported by charity. Each autumn the treasurer and a certain number of *lamas* scour the country, collecting a fee called *dubri* from each headman. The amount thus collected is considerable; and, besides that, they earn wages in the fields. Thus, during the winter, they are able to live in much comfort, and get through, I am told, vast quantities of *chang*.

One monastery is just like another. Evening was closing in as we left the *gonpa* and descended the hill. It was very chilly; but we turned once more to look at the holy flags fluttering over a building held so sacred by the followers of this curious faith—a faith which teaches that there is no God, no soul, no future state; that man is answerable but to himself alone for the misdeeds committed during his lifetime, and yet which counts its followers by millions.

These sacred prayer-flags, flags of the "windhorse," deserve a passing notice. Simply strips of white linen printed with mystic sayings, and profusely ornamented with the figure of the white horse supposed to drive away demons, they are never taken down, but flutter uncannily in shreds from monasteries and mountain-tops, houses and even trees, each breath of wind wafting another prayer to the credit of the people of the country. The well-known prayer-wheels, which date from the fifth century, are merely metal cylinders with a stick passed through, on which are prayers, printed or in manuscript. A weight attached to the centre causes the cylinder to revolve with a mere twist of the hand, and each revolution is equivalent to repeating the prayers contained in it once. Hand prayer-wheels are generally about three inches high, by two in diameter, and made in brass, copper, or silver. Sometimes they are of huge size, and turned by wind or water, and termed prayer-mills. In time, possibly, fire and intricate machinery will assist. Prayer-wheels and mills are invariably to be found in the vicinity of monasteries and temples, for the benefit of the passers by; it is not difficult to purchase them, but the owners prefer to remove the religious writing before selling them.

About the 25th of May, as we were running so short of food, I insisted on D. going out to look for some pigeons. He walked miles without coming across any, though when he was without a gun he met them by scores. During his walk he saw advancing towards him a curious procession. Along the narrow path came three wild-looking men, naked, but for some rags bound round their waists, their long dishevelled hair hanging far down their backs, and who waved large boughs of willow-tree over their heads. Behind them came a single man beating a tom tom with all his might, as if the entire success of the show depended on the amount of noise he could make. Behind him came another man, carrying what appeared to be an enormous bunch of wild flowers, stuck at the end of a pole decorated with coloured rags twisted around it. There followed a motley crowd of villagers, who, whenever they got a sign from the wild men in front, gave a single "yell," and then marched on silently till they received another sign, and yelled once more, walking all the time at the rate of a hunt. As they approached, D. held up his hand and stopped the procession. Finding one man among the crowd who could speak Hindoostani, he succeeded in eliciting the following explanation.

One day in the year is annually set apart for parading in procession through every village in the district, carrying an idol whose good offices are invoked to keep away sickness of all kinds, and this was the function he had witnessed. His informant gave him this explanation with a broad grin on his face the whole time, and impressed D. with the idea that he had not the smallest belief in the efficacy of the custom, but regarded it as a mere matter of form. The wild men, on the contrary, looked very stern and exceedingly in earnest. But then, I suppose, they were paid for the job. The idol, D. discovered, was enclosed in the bunch of flowers.

During our stay at Kailang we had become much interested in the work of the Moravian mission there, and it merits a few words of description. The object of the mission is the spread of Christianity in Central Asia, and considerable difficulty as to a base of operations having been met with when it was at first contemplated, on account of the opposition of the Russian, Chinese, and Kashmir Governments to granting a footing to missionaries within their territories, Lahoul was fixed upon, and the mission established at Kailang, its capital, in 1854. As before mentioned, the only missionary now in Lahoul is the Rev. Mr. Heyde, who has resided there for thirty-seven years. The children of the missionaries are sent home to be educated at the expense of the mission, with a view to themselves being trained for the work, and Mr. and Mrs. Heyde had now no children with them. The missionaries themselves never return home, and the life they lead could only be borne by those who have their heart in the work. Even now, after a residence of nearly half-a-century, they enjoy little in the way of luxury. The mean temperature for the year is only 44°, and in winter it frequently falls below zero. Living at an altitude of ten thousand feet above the sea, for five months in the year they are completely cut off from the world, as the passes are quite blocked with snow, and no post even is possible.

It is during the winter, however, that they find most work to do. In summer only the children of the native Christians attend school, as the *zemindars* require their children to work in the fields. But during the winter the schools are full. The three R's, sewing, and singing are taught, and religious instruction given. Large numbers of socks, stockings, gloves, cardigans, and Balaclava caps, etc., are knitted, the mission selling them and paying the children. Probably the fact of the payment, and the advantage of the well-warmed room during the long winter months, is some inducement to the Hindoo-Buddhists to send their children to a Christian school.

In ordinary years the snow lies eight to ten feet deep all over the valley, and outdoor work of any kind is impossible in winter. Mr. Heyde is a well-known Thibetan scholar, who has translated the New Testament and portions of the Old, into that language, together with many school books, and does much work with the printing press attached to the mission house. During all these years, however, I do not think that they have succeeded in making a single Lahouli convert. With the Ladakhis and Thibetans, they are more successful, and some of these families live close round the mission house. Though not a qualified doctor, Mr. Heyde gives much help and medicine to the poor, as there is no doctor in the country, and whenever I went up to his house, he always seemed to be attending to some case or other. Lahoul is, fortunately, a very healthy country. Smallpox is the great scourge, but Mr. Heyde has been successful in keeping it down by vaccinating large numbers of the population. I believe the native doctor sent as vaccinator by Government was objected to by the people, because they say he demanded some small fee, in kind, for his services, and the natives, as usual, thought they were conferring a favour on

Government by allowing themselves to be vaccinated. It is solely due to the rigorous climate that the natives enjoy the health they do, for they are filthy in their persons and habits. But the snow lies around their villages till the beginning of June, and meat killed and dried in the cold weather keeps good for months.

The mission has a branch in Bushahir, at a village called Poh, on the Shipki road, and on the immediate frontier of Thibet, but I am unable to say what success it has met with.

CHAPTER XXVI

ON the 28th of May Captain M—left again to have another try at the Bara Lacha Pass, hoping to get over it by hook or by crook, as it is a recognized highway during the season for thousands of mules and ponies, and the only difficulty to be surmounted was some two miles of snow about the summit, which could be got through at night when it is hard. The Bagha river has indeed to be crossed, backwards and forwards, three times, and difficulty is sometimes experienced for lack of bridges. These bridges have much to contend against both from nature and from man. The wood to build them has some times to be brought from as far as eight or ten marches, carried on the backs of coolies. Those which span torrents are constantly swept away by avalanches, or floods, caused by the melting of the glaciers, and those that span rivers are often broken down by the weight of the winter snow. But they suffer most of all from the depredations of the improvident native traveller. There is no wood for many days' march between Kailang in Lahoul and Leh in Ladakh. No native traveller will pay for an ounce of anything not absolutely necessary; and why, he argues, should he carry wood when a paternal Government has provided some in the shape of bridges, which it will repair when gradually, through the disappearance of the planks, they become impassable, first for animals and then for men? These depredations are difficult to stop. The road is a veritable desert, and to mount guards over the bridges would be impossible. A similar fate befalls the rest-houses or block-houses, erected for the protection of travellers from storms, so I have ceased to pity them.

Bad news continued to arrive from Kullu. The people were dying by hundreds, some on the mountains whither they had fled, some in the temples where they had rushed to implore the assistance of their gods. A friend wrote to us he had lost three servants, whom he had had to bury himself. The poor people were panic-stricken, and nothing could be done for them.

On the 1st of June, Derby Day, we moved camp up to the mission cottage, kindly lent us by Mr. Heyde, and built by him when laying out the mission farm, and engineering the irrigation works from the different glaciers. It is charmingly situated on a spur to the left of the monastery, and slightly above it, and faces the wonderful triangle formed by the Chandra and Bhaga rivers, with its apex and the Bara-Lacha Pass, a veritable field of ice. Opposite the verandah of this pretty little cottage, nestling among the pencil cedars, we could see a huge glacier, the name of which we could not ascertain, but which looked as if it must fall, if not firmly wedged in between the mountains. To the right lay the monastery, to the left the Bhaga valley, and beyond that, in full view, the Bara Baghal Pass, now seldom used, as in some places one can only cross it by being hauled up and let down by a rope. Immediately in front was the Kardang Hill, fifteen thousand feet, and towering behind and above it, a snowpeak, marked in the Ordnance Survey as "Snowy Peak M," having an altitude of 20,356 feet or three and a half miles. Surrounded on all sides by glaciers, as we were at the cottage, though the sun was hot in the middle of the day, at night the wind blew so keenly that I was glad to sit before the fire with an ulster on! We thought ourselves very lucky to be in this perfect climate during one of the most unhealthy years the Punjab has ever known.

On the 3rd of June we were honoured with another visit from the Thakur Hari Chand, who wanted D. to help him with some letters he had received from Simla with reference to the road between Lahoul and Leh. He is one of the natives entitled to the distinction of a seat in the presence of a European, so D. offered him a chair in our room, and he sat for about an hour. After he left us, he went to one of the servants' tents and sat smoking with them for a quarter of an hour before he finally departed. How incomprehensible are the ways of natives! Here was a baron, one of the highest families in Lahoul, feared and respected by all, hob-nobbing with the lowest of my servants without losing caste. How inscrutable, too, are their absurd prejudices against food, while they will drink anything in the way of water, no matter how foul. Down country, for instance, a high-caste man will throw away his meal if the shadow only of a Christian falls across it while it is being cooked. But in Kashmir I have seen them, in Srinagar, surely the filthiest city in the world, drinking the water immediately below a foul drain from the slaughter-house, carrying on its surface dead dogs and other refuse. In many parts of India they drink from stagnant pools and tanks where they also wash themselves and their clothes, or the muddy ooze in the rice fields. Yet this foul drinking in no way seems to affect their health, except in time of epidemics. There is no doubt that the rapid spread of cholera in Kullu was due to the natives all down the valley drinking of the Beas, into which many half-burned corpses were thrown.

The rainfall in Lahoul is precarious in the extreme, and the agriculturists have to depend almost entirely on irrigation. The seasons have, indeed, altered, and for the better. Thirty years ago, the annual rainfall was but a quarter of an inch, now it averages as much as four and five inches, and this year, 1892, the great quantity of eighteen or twenty inches fell. The fields are all cut out of the side of the mountain and lie on a slant, so the water is brought by aqueducts from the glaciers above, to the level of the uppermost field, whence the watercourse is zigzagged down them. Every eighth day, a party of women (women always do this work) come to the field armed with sticks, with heads like polo-sticks, and tapping the watercourse where necessary, walk backwards down the hill through the crops before the water, directing it so skilfully with the heads of their sticks into hundreds of little channels, that not a square inch of the field remains unwatered. The ground remains flooded for twenty-four hours, when the water is cut off again for another week. Where the hill side is too steep for cultivation it is irrigated for hay, so that not a drop of the precious liquid is wasted. So careful are they of it, indeed, in the Gar district, that they will continue irrigating the whole night, by torchlight. The ground but requires to be watered and protected from the cattle to produce the most delicious hay, but without irrigation grass would only grow in patches, rank and useless. Higher up the mountains, where the snows melt more gradually, is a thick growth of short grass, wherever the slope permits, affording excellent pasture for sheep, which are grazed there for the three summer months.

Lahoul is divided into three districts, Rangloi, the valley of the Chandra; Gar, the valley of the Bagha; and Pattan, which is the name of the valley after the rivers join. My remarks on the irrigation apply only to Pattan, and the lower portions of Gar and Rangloi, which are the only parts of the country which are cultivated. The district of Pattan is below eleven thousand feet, whereas the other valleys run up to thirteen thousand feet. The limit of cultivation is between twelve thousand and thirteen thousand feet; above that grass grows without cultivation to the snow line, fifteen thousand feet or thereabouts.

On the 5th of June we had snow and sleet all day; the glass standing 44° at seven a.m.; at noon, 40°; at seven p.m., 42°. June 6th was like a Christmas Day, the whole country under snow. Temperature, at seven a.m., 32°; at noon, 38°; at seven p.m., 40°. One of the servants' tents broke down during the night with the weight of the snow, and the cottage, though fairly warm, leaked like a sieve as the snow melted, and we could only keep our beds dry by moving them about whenever we saw the water coming in. The snow cleared off in the morning, and later in the day we got down to Kailang to visit the Padre Sahib, who was greatly concerned to hear of the leakage, and insisted on sending men up to repair it next day. Though Kailang looked but a stone's throw below us, there were forty-three zigzags between the cottage and the mission, and it took us three quarters of an hour to ride, and an hour to walk up.

The next mail brought news of the death from cholera of our banker in Kullu, by name Muttra Pershad. A nice man, useful to Europeans, he had made his fortune, as I have before said, by purchasing sapphires in the old days when these were brought in Kullu by the *kilta-* (as much as a wheel-barrow) load and sold for what they would fetch. But though a rich man, he lived, as the Maltese are said to live, on the smell of an oiled rag! We heard two European friends of ours had been attacked by cholera, but had recovered. The padre said there had been one or two suspicious cases in Lahoul of immigrants from Kullu, but that it was doubtful if the disease would get any hold in this cold climate, where the women are working at midday in the fields clothed in flannel and worsted socks, with a mean temperature of 42°.

We walked into the Shakur *nala*, about half a mile from the cottage. It was merely a succession of stupendous precipices forming the banks of a torrent. But this is the land of precipices, and we had ceased to be impressed by them. One in the valley of the Chandra, opposite Gundla, rises like a wall eleven thousand feet out of the river. The great depth of these *nalas* is accounted for by the nature of the soil—slate and clay—which is easily acted on by water, and results from the constant wear of countless ages. This great depth is a worry to the poor farmers, who see their lands parched on the hillside, while the rushing stream far, far below, seems to laugh at them. Mr. Heyde, however, has solved the difficulty as regards the mission farm. Lying on a spur between the Shakur and the Bailing *nalas*, he has made two aqueducts from each, and not without difficulty. At first he built a reservoir at the top of the hill, superintending the work himself; but, being called away to Ladakh before the sustaining wall was completed, he was obliged to leave his native overseer in charge. When the water was let in, the wall burst at the exact spot where Mr. Heyde had ceased himself to overlook the work. The water poured down the very steep mountain side into the village of Kailang. The people, panic-stricken, thought their last hour had come, and appealed to their idols. These, they assert, saved them; for no lives were lost, and but little damage done. Mr. Heyde, however, took the affair so much to heart, that he abandoned the reservoir scheme, and succeeded in constructing an aqueduct elsewhere. The remains of the wall are still to be seen on the hill-top, looking like an old fort, and may some day be a puzzle to the archaeologist.

The astonishing thing about Mr. Heyde's aqueduct is, to my mind, the sum for which he made it. It is three miles in length, and there was much blasting to be done. The *zemindars* round about declined to assist in what they imagined to be an impracticable undertaking. The Rhone Glacier aqueduct in Switzerland cost half a million sterling; the one from the Bailing *nala* in Lahoul, three miles long, only £33. True, there were no royalties to be paid. No covering in of the duct was necessary; no engineers were required, and each labourer was paid at the rate of less than twopence a day. Nevertheless, there were many difficulties to contend with, and credit in every way is therefore due to Mr. Heyde for his work.

CHAPTER XXVII

ABOUT this time one of our servants complained of fever. I was dosing him with quinine, when I remarked on the scantiness of his clothing, and asked how he spent his good wages. He replied that he was saving them up to buy a new wife, the probable cost of whom would be about fifty rupees. Now, he had already had three, and the most recent had lately run away from him, like the rest. I inquired the reason. He gave none. Again, I inquired if he beat them? Of course he did? was the reply. This accounts for the milk in the cocoanut. Yet here he was, half-starving and half-clothing himself, to buy another.

The Lahoulis are intensely superstitious. A few years ago there suddenly appeared in the *nala*, close to where we were encamped, a peculiar kind of wild goat, which I take, from description, to have been a *kurt*, probably chased thither by snow leopards. The whole village turned out, cornered it, and stoned it to death. Then they sold it to the missionaries, who were by no means averse to a change from everlasting mutton. Next year smallpox broke out badly among those villagers who had assisted at the demise of the goat, and many died. This misfortune was attributed to the goat having contained the spirit of a deceased *lama*, and they looked upon it as a judgment on them for having caused its death, the special demon who presides over smallpox having been sent among them. Strange it seems that they think the disease not contagious, yet capable of being conveyed from village to village by an evil spirit of wandering disposition. According to custom, therefore, each village was fenced round with a substantial hedge of prickly bushes, and similar barricades erected along the roads. Curious to relate, the disease yet spread. Perhaps there was a hole in the hedge somewhere.

Some ten years ago a woman in Kailang gave birth to an "albino," the like of which had never been seen in the country before. It was worshipped as a god, and the mother felt highly honoured. In due course, however, another child, also an albino, followed, and afterwards a third. This was too much for Lahouli superstition. Albinos were a drug in the market; gods at a discount. The *lamas*, however, consoled the woman by telling her that they were *nag* born, i.e. children of the serpent they worship, and so they are accounted to this day. They are now about ten to fourteen years old, and resemble all other albinos I have ever seen, except that they are sun-blind, though they can see perfectly at night. Is this a peculiarity with albinos?

On the 10th of June a native of Zanskar came up to us with a Zanskar pony and two pairs of ibex horns, thirty-seven and thirty-eight inches, for sale. Zanskar is a tributary of Kashmir, in the upper part of the Sooroo valley, and bordered on the north by Ladakh, and on the south by Kishtwer. Its people have but little intercourse with the outer world, which, perhaps, accounts for their independence, for this gentleman was quite angry that we did not buy. We did not want the pony, and horns we never buy, on principle. I was told he had a snow-leopard's skin somewhere in the valley for sale, but this he declined to bring up on spec. Perhaps the useless climb of two thousand feet to our camp, before breakfast, had affected his temper; anyhow we saw him no more.

The next news from Kullu was saddest of all. A dear old gentleman, father of a young planter, who had only recently joined his son, had succumbed to cholera. Death is especially sad in these remote regions. Not many months before, when we were in Kullu, an English child fell ill with croup. There is no European doctor within a hundred and fifty miles, the only ipecacuanha the parents could get was stale, and before fresh could be procured the child died, for the nearest chemist was three hundred miles off. D. read the burial service over the poor little thing, who was buried in the garden. This last cholera case in Kullu was worse. The natives were too panic-stricken to be found to dig a grave or make a coffin. A young officer in the Native Infantry, who happened to be in the valley, assisted the son to make the coffin, dig the grave, and bury the body in the garden. In the mission garden at Kailang a small plot containing ten or a dozen graves is railed apart for the cemetery.

At Sultanpore, in Kullu, there is indeed a cemetery, but as no European lives near that capital to supervise it, and as the marble cross placed there by a European over his wife has twice been broken down, it is no wonder that the residents prefer to make the graves in their own gardens where they can look after them themselves. A missionary who lost his wife in Kullu a few years since, preferred to bury her in a secluded spot where she had spent many hours sketching, rather than in the consecrated cemetery at Sultanpore. Quite recently, however, I have to relate with sorrow that this grave also has been desecrated. It was nearly opposite the resthouse at Manali and in charge of the caretaker there, and was covered with three stone slabs surmounted with a plain cross. One morning the caretaker remarked that the cross was crooked, and on going to examine it found that the slabs had been removed from the top of the grave and the ground dug up beneath them to a depth of two or more feet. We ourselves were not many miles distant at the time, and when we received news of the outrage, immediately repaired to the spot to investigate and to report the matter to the authorities with a view to instituting a searching inquiry. Captain Banon, however, who lived close by, was there before us, and had done everything that was needful. The ground was left as it was till the following day, when the chief police officer of the district arrived, all the headmen of the villages for miles round were collected and closely examined. But up to the time we left Kullu, however, no clue had been gained as to the perpetrators of the outrage. The miscreants had failed to reach the coffin, having probably been overtaken by daylight. What was their real motive must be a matter of conjecture. We inclined to the idea that they were actuated by desire of loot. The police officer, however, thought they were prompted by revenge, some of the inhabitants of the district who had been convicted of assaulting a *sahib* and breaking his little finger, having recently come out of jail. I mention this incident to throw a light on the real character of these Kullu people, so pampered by successive administrators.

I cannot leave the subject of cholera without relating a fact which has recently been told me. We have been accustomed to hear a good deal of the rascality of the heathen Chinee, but after much experience in many lands, we think we must give the palm in that respect to the Brahmins, and I am sure many readers of the following true occurrence in the year of grace 1892, will agree with me. For nearly forty years, Kullu, a particularly favoured country, had escaped a visitation of cholera. That of 1892 was probably due to the wandering tribes of Lahoul and Thibet, who winter in the plains, returning to their own country in the spring through Kullu. It was of exceptional severity, and hardly a village in the country escaped. After such a long immunity, the inhabitants were panic-stricken, and fled to the mountains and forests to escape its ravages. There happened at this time to visit the country a Brahmin from Benares—Benares the sacred city of Hindooism, the holy of holies, laved by the blessed waters of the Ganges. From such an odour of sanctity, and at this time, came this Brahmin, who saw his chance, and promptly laid himself out to profit by it. He perceived himself to be among a simple uneducated race, panic-stricken under the iron grip of the scourge, and he proceeded to work upon their fears and credulity. Taking unto himself two other scoundrels of the same religion, the three planned to make a tour of the cholera-stricken villages, he representing the native doctor, one of the others his servant, and the third the native police officer, and pretending to have been sent by Government to make a sanitary inspection. Such a sham inspection they carried out, and the people were told that the only chance of saving their lives was to flee to the mountains. The false police officer announced that he was instructed to see the orders of the native doctor carried out, and the villagers, who dread the police officer more than any earthly power, fled *en masse* as they were bid. As soon as the coast was clear, the three scoundrels made a house to house visitation, looting every thing, and ransacking every box and cupboard. But, as luck would have it, they did it once too often. In one village a boy had accidentally been left behind, who concealed himself and watched their depredations, and who informed the inhabitants directly the thieves had gone. The people hurried back, caught the wretches redhanded in another village, and laid information which led to the arrest of the sham doctor and his servant. The false policeman, I regret to say, is still at large. The holy Brahmin was awarded six years' rigorous imprisonment, his coadjutor the same, with the addition of periodical public floggings by a sweeper, the lowest caste in India.

Early on the 12th of June our attention was attracted to the monastery by an unusual tomtoming and shouting. On inquiry we found that the contents of the library were being returned by a neighbouring farmer to whom they had paid a visit. We saw them being carried up the hill in *kiltas* (cone-shaped baskets), on the backs of laughing and joking coolies, who receive no money payment for their precious loads, but are plied with as much *loogri* as they can carry. Such a visit from the one hundred and eight sacred volumes is supposed to ensure prosperity to the household whither they are taken. For two or three days the *lamas* sit reading them constantly. The *zemindar* supplies the *loogri*, and before dusk on the day of their return to the monastery, men, women, and *lamas*, are all as drunk as they could be.

This is the sort of orgie countenanced by the Buddhist clergy! At Lhassa (the abode of the gods), and the capital of Thibet, I am told morality is at an even lower ebb. Lhassa is the Buddhist Benares, Rome, or Mecca. It is strictly closed to Europeans, only one Englishman, Manning, having ever penetrated within its walls. The depraved *lamas*, believing in no future state, naturally like to keep themselves to themselves, and dread the cleansing influence of civilization. There are two sects of Buddhists, the red—Drukhpas—those of Lahoul; the yellow—Gelukpas—those of Lhassa and Thibet. The latter, celibates, are supposed to practice a purer form of faith, but from evidence of travellers they are morally much inferior to the red sect. One would expect to find the clergy, the only educated people in the country, superior to the laity. But such is not the case. The latter have many good traits, are fairly truthful, and quite honest. Theft is almost unknown; so is murder. Their drunkenness must be attributed to their climate, and to the evil example of their spiritual advisers, who are also accountable, in a large measure, for their immorality, the natural outcome of the celibacy of a large proportion of the population.

CHAPTER XXVIII

ON the 13th of June we started to visit the Shakur glacier, but after an ascent of two thousand feet we came to the conclusion that we were near enough, and that we should not see it to greater advantage did we go up it. We were mistaken, however, and we made another ascent later on. From the spot where we halted we had an interrupted view of no less than fourteen glaciers—a truly magnificent spectacle. D.'s aneroid marked 18.50, or just over fourteen thousand feet above the sea level. It was piercingly cold, and we were soon glad to beat a retreat. At the very spot where we sat, nestled beneath a rock, we found a fern, a great rarity at this height and in this country, where they are almost unknown. We took it down, and also much edelweiss and some forget-me-not. Doubtless, later on, the spot would have been carpeted with wild flowers. We had left the pencil cedar forest far below us, and it was curious to mark its upper line, almost as clearly defined as the snow-line. I scanned the horizon with the glasses vainly for ibex, but we saw some snow pheasants (*golind*) in the distance.

This is the home of the snow-leopard (*Felis uncia*), that rare and beautiful beast which most sportsmen would sacrifice their dearest hopes to obtain. But we found it as hard to get a specimen as to get a shot at a panther in Kullu. Since we had got into Lahoul, several ponies, donkeys, and sheep had been killed by snow-leopards; and though we offered handsome rewards to any one who would bring us news of them, we only succeeded in hearing indirectly of kills some days after the event. Two explanations are given. The first is that the shepherd is generally alone with his flock when it is attacked, and were he to leave it to carry news to the *sahib*, who knows what havoc would not be done in his absence? The second is that, were the *sahib* to miss, the leopard might charge and the shepherd might be wounded; and, as they are only servants in charge of the flocks, they prefer to lose a sheep or a goat to running the risk of losing their own lives. Leopards are nasty beasts to deal with, as any one who has had anything to do with them knows. Only recently I read an account in the paper of a panther down country, who had accounted for a hundred and fifty-four human beings before he himself was killed, and it was undoubtedly a fact.

About this time the Thakur sent us word that the snow-leopards had been at work around his village, where they had destroyed some of his ponies, and begged us to come and put an end to them. But the invitation did not tempt us. It was a twenty-four mile ride, and probably a ten mile walk at the end of it. Our experience told us that the only chance of a shot at a leopard is when he himself is engaged in a stalk, and as the chance of that was a thousand to one, we preferred to wait for a closer opportunity.

In these regions the rain, instead of coming down, comes up, in torrents. On the 14th of June we saw the clouds far below us long ere the rain reached us, and when it did come it was snow, and the mountain sides were once more covered. A thunderstorm up at this height is a magnificent, awe-inspiring spectacle. The clouds ascend, each flash of lightning gives a momentary glimpse of the surrounding snowpeaks, which appear grander than ever when the darkness of night is torn from them, and they are illumined suddenly by the fierce light. The thunder follows, a single clap, but echoed and re-echoed from valley to valley and peak to peak, until another flash comes and another clap of thunder, so that a continuous roar is kept up till the storm has passed.

After the tempest, we looked down on a rainbow, far away in the valley below the Bara Baghal Pass, the colours coming out most beautifully. We came across a "marmot" when out walking that day. A little tail-less animal, the size of a small rabbit, dashed across the path into a hole in the rock, so quickly that I had not time to see what it was, and there are no rabbits in Lahoul. This marmot is unlike the Kashmir species, and the smallest kind extant. Mr. Heyde, who had seen them close, described their fur as of a beautiful reddish brown, and esteemed of great value. We afterwards caught a couple of these beautiful little creatures, had a box made for them, and kept them until we left the country, when, fearing that they would not stand the heat, we let them go. They were the dearest little pets imaginable, and the tamest wild animals I have ever seen. They fed entirely on grasses and wild herbs, and were so unlike rats in every way that I got D. to write to the Natural History Society, India, to ask what the real name for them was. We ascertained, through the society, that they were not "marmots," but "mouse hares" (*Lagomyo ladaccensis*).

The colour of the water in Lahoul puzzled me much. The water of the glacier in front of us emerges from it a dirty black colour, showing that the glacier rests on a formation of slate. In the Bailing *nala*, however, it is exactly like milk, denoting a foundation of granite, and is much valued for irrigation. In both cases, however, the deposit left acts instead of manure. In Switzerland farmers pay more for water which flows over a granite formation. Now I was puzzled that the water should grow clearer the further it flowed from the glacier, and have built up the following theory to account for it. I fancy, under some glaciers, at all events, lies a sea of mud, which colours the water flowing over it in its downward course beneath the glacier; and, as the stream is fed by other streams of clear water, it gradually loses its muddy appearance. A village, seven miles up the valley from Kailang, was completely overwhelmed by a mud avalanche from beneath the glacier of Deeno, though the glacier itself did not move. It was a unique occurrence, and the village has been rebuilt and re-inhabited.

The inhabitants of Lahoul drink of this discoloured water with impunity, and yet goitre is unknown. Snow-water is their only supply. In Kullu, on the other hand, where the water is abundant, and pure and clear as crystal, one out of every three or four of the inhabitants suffers from this disfiguring malady. But in Lahoul there is little vegetation, and Kullu is nothing but forest, and is remarkable for its rank vegetation. So, surely, goitre is not due to the indiscriminate drinking of snow-water, but of that heavily laden with decayed vegetable matter.

We inspected one day a Lahouli house, like all the rest, low, flat-roofed, and full of posts to support the roof, for lack of long beams. We were surprised and amazed to find in it a glass window and a stove with a chimney. As soon should I have expected a white-capped English parlour-maid to open the door. It was the usual explanation—the Padre Sahib. What a beneficial influence has that man in Kailang! As the Lahoulis are confined to their houses during long winter months for weeks together by the severity of the weather, what a boon to them must be that glass window, replacing the holes in the roof, where the snow comes down, and the foetid, blinding, smarting smoke of the cow-dung fire does not get out. What a blessing that stove and its chimney! In front of the house a *Dzo*, the hybrid *yāk* and cow, was being milked, its calf as usual (in this case a year old and as big as its mother) tied up before it. All over India exists this extraordinary notion that a cow will not give milk unless its calf is near it. Last winter, when I had a cow, I gave away the calf, and got no milk in consequence. I begged the calf back, and the milk returned, but to vanish when the calf shortly after died. Now I can milk a cow myself, and, of course, the cowman, and not the cow, was at the bottom of the trick.

Red and yellow roses were now in bloom everywhere, the former the ordinary English dog-rose, the latter, I think, a kind of Austrian briar, a superior double rose, sweet-scented, and glorious in colour. But, on Waterloo Day, again the hills were covered with snow.

We were again horribly annoyed with the natives. The coolie who gathered wood for us informed us, in the most nonchalant way, that two nights since a snow-leopard had killed a sheep in his village, half a mile below us, quite early in the evening, and that the people sat on the house-tops watching it. Had he but brought us the news, we could have run down the hill in no time and should probably have got a shot. Later on, when Mr. Heyde came up to see us, he gave us another explanation of this reluctance to give news of a snow-leopard, viz. pure superstition. The Lahoulis call the animals "semchun," which, in Thibetan, means, "a living being," instead of "shun," which means "snow-leopard;" and they imagine it contains the spirit of a departed human being, and so are afraid to kill it, or to be accessory to its death. Mr. Heyde told us a few years back that the father of the present Thakur, when accompanying some high European official on a shooting trip, was horror-stricken because an ibex was killed, but consoled himself by pointing to the official and saying, "Half the sin will rest on my soul and half on his!"

On the 19th we went up the Bailing *nala*, in hopes that the two days' snow might have driven some ibex down. On the way up we crossed a side *nala* over snow ten feet deep. On our return we found it had been carried away *en masse* by a mud avalanche, through the remains of which Dick had to wade across in eighteen inches of mud, though I succeeded in getting a coolie to carry me over. The mud avalanches have caused much trouble from time to time to Mr. Heyde's water works. Coming down with great force, they cut deep into the *nala* and left the head of his aqueduct high and dry, till, at last, he found a place where the bed of the *nala* was solid rock, and there began his duct. Three days after the avalanche, the water issuing from the *nala* was still as black as ink, proving to my mind the existence of a mud sea under that glacier at least.

CHAPTER XXIX

THE *guddis* now arrived with their flocks. We came across them one morning, a hardy, weather-beaten looking race—and could not at first imagine what they were about, lying flat on their stomachs with their faces to the ground. We discovered that they were smoking. They had put a little tobacco into a hole in the ground, placed wood ashes over it, covered with wet clay, with two holes in it through which they inhaled the smoke. It must have been warm, if nothing else. At the junction of two roads, in Kullu, I came once across a public pipe. It was made of clay in the shape of a pear, twelve to fifteen inches in circumference, the thin part uppermost, and a piece cut off to make two holes at the top. The smokers also had to lie on their stomachs to enjoy it.

Mr. Heyde frequently regaled us with tales of the manners and customs and superstitions of the *guddis* and the Lahoulis, which are very quaint. When they kill a sheep or goat, he has seen them tear it to pieces with their hands, and eat the flesh raw. This habit probably arises from the fact that, when travelling in Ladakh, they cannot carry or find firewood, and cannot cook their food. Another custom of the *guddis* is almost too loathsome for description. They puncture a hole in the side of the goats, near the kidneys, and inserting a straw drink the fat! The custom is not at all uncommon.

One of the Thakurs recently wished to build a shop for the disposal of his farm produce, by the roadside, near the rest-house at Kailang. Before commencing, however, according to custom, he sent to a *lama* to ask if the undertaking would be propitious. The *lama* came to the spot, carried a handful of earth to the monastery, and after three days returned to say that it was not propitious, whereupon the idea was at once abandoned. According to this the great men of the country would appear to be entirely in the hands of the clergy. But a give-and-take principle prevails. Should the *lamas*, after consultation with the books and stars, fix a certain feast day, or even the New Year, which varies also, for a certain date, the Thakurs will drive a coach and four through their decisions, and revoke them. I should not be surprised to see in the Lahoul Gazette, if there was one, that even the New Year had been indefinitely postponed!

When we went down to the post-office next mail day to ask for letters, we were utterly amazed to find one of these wild-looking *guddis* asking the Babu to write a letter for him. One would as soon have expected to find one with a pair of boot-trees. Many years ago in Trinidad, in the West Indies, D. remembers how troops of emancipated niggers, thick-lipped and woolly-headed, who had been born in the island and had never left it, would troop down to the post-office when the English mail came in, in top-hats and patent leather boots, and inquire for letters from home!

Simultaneously with the *guddis* arrived Thibetan traders from Rupshu, a tributary of Ladakh lying to the south-west of the Indus, bringing with them a hundred and fifty Biangi sheep, peculiar to their own country, laden with salt. This they exchanged for corn—five measures of corn for four of salt. The wool they sold for one shilling and threepence halfpenny for every two fleeces, but only to purchasers who would give them corn for their salt. Mr. Heyde bought all the hundred and fifty fleeces. But there were not sufficient for the requirements of the mission, which last year turned out over eleven hundred pairs of woollen stockings, knitted by the school girls. In the afternoon we went to inspect the shearing. All the sheep were huddled together in a pen, and under a shed sat a dozen Lahoulis engaged in shearing Shearing sheep in Lahoul them. In one corner of the shed sat the owners of the sheep, of pure Chinese type, and wearing pigtails. They were not asked to help to shear, as they do it with their knives less dexterously than the Lahoulis with their shears.

So they sat taking no notice whatever of what was going on, round a small fire, on which was a cauldron containing the concoction of tea and melted butter, so much drunk in these parts. In the top of the cauldron was a large ladle, in each man's hand a wooden cup. In the middle a bag of barley flour, which they sometimes mixed with tea into a paste, and sometimes dexterously threw by pinches into their mouth and washed down with tea. This flour is made from grain that has been previously baked, after the manner of infants' food at home, and is said to be very wholesome and digestible.

None of these Rupshu traders, or their forefathers or foremothers, for generations back, have ever possessed a house. During the summer they live with their flocks on the high lands of Thibet, with a preference for Rupshu; in winter they repair to the valleys of the Indus and the Sutlej. Naturally a hardy race, and used to cold, it is with difficulty that they are induced to come to Kailang, for they fear the heat, though the place lies ten thousand five hundred feet above the sea, and the mean temperature this June, when we had two falls of snow, was 49°.

When I admired the fleeces, I was told that the Chumarti sheep were much larger—as big as donkeys—and cost fifteen shillings each. Here the price of a sheep is one shilling and sixpence to two shillings. I ventured, through the interpreter, to mention that in England sheep cost from three to four pounds apiece, but this statement was received with laughter loud and long; and I was told that Lahoulis were not such fools as they looked, or words to that effect. They thought I was drawing the long bow.

The sheep which had looked so handsome with their fleeces on, presented a wretchedly meagre appearance after shearing. They were so thin that you could almost see through them, and the patch of wool left in the middle of their backs as an offering to the *lamas*, and the patches on the back of their hind legs, to prevent their being cut by the rope that hangs over their stern to keep the loads in place when they return laden with corn, gave them a very ludicrous look. But I was told that this year, owing to the mildness of the winter, they were in exceptionally good condition. What must they be after a severe winter!

CHAPTER XXX

WE were lucky enough during our stay at Kailang to come in for the event of the year in Lahoul, the annual *lama* dance or *cham*. Like most oriental *tamashas*, however, such as the dancing dervishes of Cairo, or the howling dervishes at Smyrna, or the devil-dancers of Ceylon, if you have seen them once, you do not wish to see them again. All Buddhist festivals are movable feasts. Before, therefore, the date of the *lama* dance could be settled, the gods had to be consulted. It generally takes place in the middle of June, but this year it was postponed on account of the cholera in Kullu, and of the fear that, by congregating large numbers of people to gether, the disease might be imported into Lahoul. As this country appeared, however, likely to escape the scourge, the *lamas* decided that the 9th of July would be the most propitious day for their annual festival. The gods were intrusted primarily with the responsibility of fixing the date; but the Thakurs had also much to say to it, and they decided, very wisely, we thought, upon the day before the full moon, for the orgie after the dance is prolonged far into the night; and there is no doubt that not a few of the assistants require all the moonlight they can get to find their way home in the small hours of the morning.

For three or four days before the great event, whenever we went down the hill from the mission cottage to Kailang, we met *lamas* from distant parts, converging on the monastery to take part in the approaching festival. Quaint-looking men they were, in their sombre dress, bearing their prayer-wheels always at work, but filthy to a degree, and the privilege of meeting them on a narrow pathway was a dubious one. D. says that there is a bouquet in Africa, known as *bouquet d'Afrique*; the scent very searching and pungent—he always thought it—and one of the most powerful scents known to man. But he declared that it could not compare to *bouquet de lama*, the true pachouli of the East!

Sundry discordant sounds, emitted by unnameable instruments, ushered in the dawn of July 9th; and a glance in the direction of the monastery showed us that extensive preparations were being carried out there. An increased number of the white or sacred flags, a few coloured banners, red and white, green and white, or blue and white, innumerable rags, the size of a small pocket-handkerchief, dyed red and yellow, and tied to a string, hung from different corners of the building. On the left-hand side of the door leading into the *gonpa* were suspended huge paintings on silk of Buddha, and another adorned the "may-pole" in the centre of the dancing-green. As they passed these painted figures, I noticed that the women all *salāmed*, but that the men did not. The smoke issuing from the courtyard of the monastery indicated, probably, the brewing of *loogri* for consumption at the conclusion of the dance. Very early in the day the sounds of singing—if singing it could be called—announced the arrival of different parties of the Lahoul ladies. We watched them streaming along the mountain-paths leading to the monastery, shouting and laughing as though the word "care" were an unknown one in the Thibetan language. They have very pretty feet; and I noticed that as they neared the *gonpa*, their grass-shoes were replaced by leather ones, much in the same way as a barefoot Irish girl dons her shoes as she approaches the town on a fair day. All natives, however, are particularly partial to appearing in public in shoes. Though accustomed to go barefoot, and being blest with soles as callous as a horse's hoofs, they save up money to buy shoes to wear on great occasions; and many a household servant will stint himself of his pay to purchase socks in which to adorn his feet when on duty, though it is absolutely unnecessary. Some of the fair visitors carried umbrellas—a fashion less common here than down country. Many years ago such luxuries were unknown, or, at all events, untolerated, by the lower classes of India. But only recently I read in an Indian paper how a rajah near Calcutta had celebrated his birth day by presenting several thousands of umbrellas to his poorer subjects. Not long since, if a native carrying an umbrella met a European, he put it down while passing him, as a mark of respect. But "nous avons changé tout cela." Thanks to Lord Ripon and his colleagues, one man is as good as another, and a good deal better!

As we sat watching the arrivals there came up to us the *Nono* of Lahoul, the adopted son of the Thakur, Devi Chand, of whom I shall have something to say later on. He had come to show me a tiny snow-leopard. Its mother had killed some sheep above the Thakur's village, and the Nono had gone out to shoot her. When he reached the cave where she was supposed to be, he told us that he saw a long white thing lying across the entrance, which he took to be the leopard, and fired. He then saw one little mite jump up and run into the cave, and when he approached he found that he had shot the other. They were lying facing one another, and he mistook them for one animal. The mother was away, so he had no difficulty in catching the live one, and after waiting some time in vain for its parent's return, he carried it home, and next day brought it to me. It was, apparently, about a fortnight old, and a beautiful little creature. It had a most extraordinary cry, like the shriek of a parrot, which no doubt on a still night could be heard at least a quarter of a mile away. Its fore paws were already very powerful for its size; it was slightly underhung, and its jaws very strong. Some day doubtless it would develope into a veritable *budmash.** There was a great difficulty in feeding it. We tried it with milk in a sponge, with a bit of indiarubber tubing, and with a hole cut in the thumb of a glove. It took kindly to none of these make-shifts. My pet goat, however, had recently had a kid, and we gave her the little leopard to nurse. Moti, however, declined to lie down to suckle him, and he found it inconvenient to stand up to feed, and commenced to shriek so discordantly that the goat grew frightened and began to kick freely, which put an end to this plan also. And at last I found a teaspoon most successful. Now, as these snowleopards are very valuable, and as I do not think there are any in England, we wished sorely to

* A very expressive Hindoostani word for a "bad character."

rear him and take him home. Mr. M—, however, down in Kullu, who had lent the *Nono* a rifle to shoot snow-leopards with, had first choice.

Contrary to our fears, it turned out quite strong and healthy, for we hardly thought it possible that he would survive the change from the snows of Lahoul to Kullu in the hot weather. He fell into good hands, and was at once deprived of his ulster coat, being shaved to the skin. He much appreciated the riddance, recovered from his indisposition, the result of the change of climate,

My Pet Snow-Leopard Cub

and at once became the family pet. When we saw him again, three months later, he had grown his coat and was an exceedingly graceful little beast, sleeping on the Mem Sahib's bed and answering to the name of Moti (pearl) she had given him. He followed her everywhere, even running along the road on a fifteen-mile march between Katrain and Manali. But he was already developing the natural ferocity and treacherousness of his race, and learning to make pretty free with his claws. We gave him a wide berth, except when tired out with a long march, and disinclined to mischief; for he had left the marks of his poisonous natural weapons on the hands of his master and even of his mistress. Before we bade him a final adieu we weighed him, and he turned the scale at twenty pounds; and I was photographed with my old pet in my arms. But he had already bitten a boy, and had taken to knocking down goats, so that in a few months we foresaw that it would be necessary to consign him for life to an iron cage, or sell him to some rajah with a fancy for a menagerie.

The arrival of the new pet and the attempts to feed him had occupied some time, and the *Nono* now suggested that we should accompany him to the monastery. But it was still too early, and eventually we met him there about two p.m. The *Nono* conducted us to a blue carpet near the centre of the dancing lawn, upon which we seated ourselves. I noticed that the *lamas* present—there must have been sixty or eighty of them—were dressed in festal red instead of the usual brown garments. This showed that they belong to the Drukhpa or red, as opposed to the Gelukpa or yellow sect of Buddhists. The red was of every tone of dirt, but I must say that it was exceedingly picturesque.

Immediately on our arrival four *lamas* emerged from the monastery, two with small trumpets, and two with trumpets at least eight feet long, and all perfectly straight. The smaller ones sounded two notes; the larger, which could only be blown when the end rested on a wall, but one. After two or three blasts "on this strange horn," much after the manner in which the Wagner plays are preluded at the Baireuth festivals, the band appeared.

It consisted of the two principal *lamas*, of two old men, and two boys. The monks wore red, and huge, cardinal hats, the latter were bareheaded. The former performed upon the cymbals, and most dexterously they did it. I had no idea that so many shakes and quivers could be got out of such simple instruments. The others played the tom-toms. These were not the common Indian tom-toms, but instruments shaped like a flat circular footstool and about the same size, and stuck on the ends of thick sticks three feet long, held in the left hand, the ends resting on the ground. The drum-sticks were shaped exactly like a note of interrogation, and were held at the lower end, while at the upper, rag was bound round to form a head. Unlike the cymbal players, the drummers showed no proficiency in their performance. The band marched down the steps leading from the *gonpa*, each member curvetting and pirouetting all over the place, till these reached the seats provided for them near the centre of the dancing green, where they placed themselves, the *lamas* in the middle, and the drummers on either side them. After a few moments the overture ceased. A monk, standing on the top of the steps, put his fingers in his mouth and gave a shrill whistle, answering to the "call bell" of our theatre, and, there being no curtain to rise, the play commenced.

It was a matter of much regret to me that no one could explain the meaning of the mystery play to us, except in the Thibetan language, which we neither of us understood. The *Nono*, indeed, a young fellow of twenty-one, or thereabouts, could speak both Hindustani and Thibetan, but he did not understand the meaning of the play, though he told us his uncle could have explained it if he had been present. I am therefore only able to give a general idea of what took place. The following extract, however, will throw a light on the subject:—

"The *cham* or religious dances performed in the Thibetan monasteries are worth seeing. If introduced into a Christmas pantomime in London, they would be effective as a spectacle. The abbot and superior monks, in full canonicals, sit round the courtyard of the monastery, clanking huge cymbals to a slow time or measure. Bands of other monks dressed in brilliant silk robes with hideous masks or extraordinary head-dresses, and with strange weapons in their hands, dance in time to the measure, advancing and retreating, turning and whirling, with strange studied steps and gestures. The story of the ballet is that of 'Lugon Jyabo dung-dri,' the expulsion of the prince of devils. The latter had become too powerful, and tyrannical over mankind, so the gods descended from heaven, took the shape of strange beasts, and in that guise fought with and destroyed the demons."

The origin of the play, or *lama* dance, which dates from the fourteenth century, is thus accounted for. There lived then at Lhassa, "the abode of the gods," a king named Lungha Dharma, who persecuted the Buddhists in every conceivable manner, and did his utmost to exterminate them in the country.

A *lama* who had been turned out of a monastery close to Lhassa, vowed vengeance against the king, and circumvented him in the following manner. He procured a black pony and whitewashed it. Then, taking a bow and arrow, he rode to Lhassa, where he happened to find the king in the public market place, in the act of reading a proclamation from the emperor of China. He drew his bow and shot the king in the back and killed him, and then galloped away in the direction of the river. In crossing it the whitewash disappeared from the pony, who became black again. When he was overtaken, not being known to his pursuers, he was unrecognized, and escaped, as they were in search of the rider of a white pony. After the death of King Lungha Dharma, Buddhism nourished once more, and ever since then, the festival known as the "Lugon Jyabo dung-dri," has been annually kept to celebrate his death. The Buddhists believe that he was the prince of demons, who had been sent into the world to destroy them. Hence their joy at having got the better of him.

Year after year the same performance is gone through, with the same scenery, the same music, the same masks, and, approximately, the same performers. The people are induced to witness the play by being told that the demons who appear in the second act are facsimiles of those that dwell in one of the hells through which all must pass, prior to being born again, and that it is their duty to come to the play in order to become better acquainted with the devils, so as to fear them less when the inevitable meeting takes place. Do the people flock because they believe, or because of the orgie which they know will invariably follow the dance? I fancy the latter is the true motive, for they never refer to it as a religious gathering, but always as the *mela* or fair.

The play consisted of five acts, each lasting half an hour, with intervals of a few minutes, and took three hours in all to perform. During the whole of this time not a word was spoken by the actors, but at intervals some prayers were recited by the *lama* cymbal players, which were repeated in chorus by the performers.

Act I. Boys dressed in tight white drawers and vests, and wearing death's-head masks, made their appearance. They did nothing in particular but rush about, playing the fool with one another, much like pantaloons in an English pantomime. I think they were intended to be skeletons; if so, they were very lively specimens!

Act II. Twenty or more *lamas*, robed in flowing garments of scarlet homespun, and disguised in hideous masks representing the heads of lions, bulls, and many-mouthed demons, entered, dancing and gesticulating freely. These masks, we ascertained, were all home-made, and showed considerable ingenuity. The makers would, I feel sure, drive a roaring trade in Rome during the Carnival.

Act III. Fifteen *lamas* came on, attired in garments of imitation *kincob*, of lovely colours, which, if real, would represent a king's ransom. On their heads were Dolly Varden shaped hats, the brim trimmed with black sheepskin, and ornamented with an imitation cockscomb. Coloured silk veils hung down to their waist behind. As these *lamas* are nearly all heavily bearded and moustached, the effect of these hats and veils was ludicrous in the extreme, though an attempt was, indeed, made to conceal the beard by a dirty rag tied round the neck. A *cummerbund* girdled their flowing robes, which was ornamented at the back by a silver filigree box, containing probably a paper on which was inscribed hundreds or even thousands of times the *mani dor* prayer. These boxes were lovely works of art, and I tried to buy one. Their picturesque home-made blucher boots of many-hued wool, with felt soles, were tied under the knee with a coloured cord. As in this act the *lamas* appear for the first time in gorgeous garb, we took it to symbolize the descent of the gods upon the scene of action. Presently some cinerated human bones were carried in a white cloth on to the green by a couple of half-naked boys, and deposited in front of the image of Buddha which was suspended from the may-pole. These are supposed to be the remains of Lugon, the prince of demons. The youths laughed freely, and made a great joke of the whole thing; but the dancers, however, made a great fuss over the relics, cursing them loud and long, execrating them in chorus with the head-*lama*, and going through all manner of horrible contortions; waving wildly their *dorjés*, or imitation thunderbolts, to keep away demons, and flourishing in frenzy their *phurpas*, or *dorjé*-headed daggers, to destroy the same.

Acts IV. and V. To my inexperienced eye the players were dressed, and acted in these, in exactly the same way as in Act III. The only variation was the distribution from a copper teapot of much *loogri*, served in tiny copper cups, which some emptied, while others threw the liquor on the ground.

After the final act the band once more pirouetted round the dancing-green, reascended the steps, and were dismissed by a final salvo from the curious old trumpets into the *gonpa* again. What especially struck me during the entertainment was the entire absence of any applause whatever. During the whole three hours, not a bravo, not a clap. Certainly the Lahoulis *s'amusent tristement*! Yet I am convinced that they were not only pleased, but delighted, with the performance; but no one who did not understand their character would have even guessed it.

During the *entr'actes* we had ample time to take stock of the *mis en scène* and of the audience. The scenery was magnificent. The Ober Ammergau Passion-play is superbly "set" among the Bavarian Alps; but the spectacle was nothing to this. Immediately opposite us rose the side of a mountain fifteen thousand feet above the sea; two thousand feet below us lay the valley of the Bagha, with its grand precipices of grey rocks, with cultivated green patches; on either side stretched huge glaciers. We sat under the shadow of the quaint old monastery, gay with many-coloured sacred flags, amid groups of Mongolian spectators seated beneath the shade of two splendid *peepul* trees, sheltering them from the rays of a brilliant sun, which lit up the green lawn and the curious dresses of the dancers—forming altogether a picture which I shall never forget.

We had been placed, I presume, in the seat of honour, for we squatted on the blue carpet immediately behind the band. The *Nono* sat beside us, and amused himself nearly the whole time by looking through Dick's field-glasses. It seems that he had purchased a long telescope, but did not think much of it. On inquiry, I found that one of its glasses was broken, but he did not appear to think that that mattered, as there were plenty more. The *Nono* was the most important personage present, and ordered every one about, and was promptly obeyed. He even chaffed the head-*lama*, telling him to make haste, to play louder, etc. I believe he would have chaffed the Archbishop of Canterbury at a confirmation, this irreverent young man.

As the men were now away trading in Rupshu, women predominated among the audience. Every village for miles round had sent its fairest, and as there are no foolish *purdah* restrictions in Lahoul, we had a good view of the beauty of the country. They wore the usual dress of brown or chocolate-coloured homespun, girdled at the waist, and pyjamas of the same hue. Their hair was profusely plaited all over their head, and in the case of married women surmounted by the *kirkitzee*, a silver cup-shaped ornament, gilt inside, and sometimes studded with turquoises. Their poor ears were borne down by the weight and number of the ornaments fixed in them, and many wore nose-rings of fantastic shape. As a whole, the women were certainly not handsome, but quite nice-looking. Our washerwoman, the blacksmith's daughter, carried off the palm among the Lahoulis. A great many of the ladies wore a green shawl, which seemed to be a very favourite colour, I suppose because it is so scarce in this country. The nuns, in the case of the younger ones, looked quite nice in their plain grey homespun, with their white shawls, red caps, and close-cut hair. But the dress is hardly becoming to the aged. A word of tribute must be paid to the Lahouli teeth, which are beautiful, so short, even, and white.

Among the spectators was a group of women from far-away Bushahir, in the upper Sutlej valley, ten marches from Kailang, and who were easily distinguishable by their flat black-cloth caps, and also by their superior looks. Their husbands were away trading in Rupshu, and had left them here till their return. Their caps were prettily decorated with flowers, and they were literally loaded with ornaments. One lady in particular we noticed, who was quite a beauty, and received marked attention from every one, from the *Nono* downwards. I admired some of the ornaments worn by the different women, and only had to mention the fact to the young potentate for him immediately to send his servant to the person we designated, with, I suppose, his compliments, and the request that the wearer would take them off and send them to us. This was instantly done, and we examined them carefully. We even opened the talisman and took out the prayers and formularies which they contained, the owners raising no objection. True that we found no photographs, locks of hair, dried flowers, or love-letters, or the case might have been different.

I especially admired a prayer-wheel which an old *lama* was turning, and the *Nono* immediately sent for it. Alas! when it was handed to me, I commenced, by mistake, to turn it the wrong way, and the look on the face of that poor old *lama* I shall never forget! In the Buddhist mind, to turn a prayer-wheel the wrong way is nothing short of sacrilege. Such an action unsays all the prayers which have been said for an indefinite period beforehand, so that poor old man would have much lee-way to make up, owing to my mischance, and his face showed that he was aware of it.

Except the ubiquitous *loogri*-sellers, in all this concourse of people we noticed but one itinerant vendor, and that man sold sugar-stick and parched corn. The latter was in great demand. The dexterity with which the women picked up grain by grain and threw it into their mouths was most remarkable. They would rival an old barndoor hen over picking up grains of wheat.

By five o'clock the dance was over, the show ended. Clad in their ordinary attire, the *lamas* emerged from the monastery, and mixed among the spectators under the *peepul* trees. The serious part of the entertainment, i.e. the drinking of *loogri*, was commenced. Men and women drank together, and as far as we could see, drank fair. We, however, discreetly retired, but we heard sounds of revelry far into the night.

When we got home I found my little leopard still asleep, but on hearing me he woke up. He was never happy unless in my lap or arms, and would cry bitterly if I left the room without taking him with me. At night he slept in bed with me, as quiet as a mouse, liking the warmth, and, strange to say, slept soundly all night, though he sometimes dug his claws into me, which was rather trying.

CHAPTER XXXI

ON the 22nd of August, having been delayed for weeks at the mission cottage by bad weather, we started for the Bara Lacha Pass. The first march was to Kolang, about ten and a half miles. The road lies always at least five hundred feet above the Bagha river. In many places it has been terraced out of the solid rock, and the precipices are grand, rather too grand for the nervous pony I was riding, especially considering the narrowness of the path and the entire absence of any sort of parapet. I took the precaution always to dismount when crossing the small wooden bridges which spanned the many mountain torrents which we crossed, some of them roaring through chasms cut deep by the rush of melting snow during the few warm months. If one of these bridges had been carried away, there would have been nothing for it but to wait till it had been repaired, or to retrace our steps.

Eight hundred feet above our first camp, in the middle of a small well-cultivated plateau, stands the castle of Thakur Hari Chand, the "king" of Lahoul. It was the usual style of Lahouli residence, and was a perfect rabbit-warren of small dark rooms. It was built in flats, and the Thakur proudly showed us his latest improvements. These consisted chiefly of a series of cook houses, arranged one above the other, with a hole in each ceiling to allow of the smoke to escape. The cooks in the topmost compartment can hardly have had a pleasant time of it.

After being shown all round we were taken to a "hall of audience," some twenty feet square, and open on one side to the air. In one corner was spread a cotton carpet, on which were placed three seats, one an English chair, the other two empty wine-cases, standing on end and covered with a bit of carpet.

Here we sat for about a quarter of an hour, talking to the Thakur, while a half-naked servant brought in dried apricots and currants on a brass dish, and laid them at our feet. We were then invited to inspect the private chapel, the only clean place in the castle. It was profusely painted in the native style, and the altar was crowded with gods and goddesses of every size and shape—Buddha, Triloknath, and Devi Tara, having the place of honour. The Thakur showed us an excellent *drilbu* (sacred bell), *dorjé* (imitation thunderbolt), and a very fine antique *phurpa*, or three-sided dagger with a *dorjé* for a handle, which was the best specimen I have ever seen, and which I should dearly have liked to annex. In one corner of the chapel stood a huge prayer-wheel, eight feet by four. It was literally covered outside with what I imagine were sacred writings, and the Thakur assured us was fitted interiorly with manuscript prayers. At the top of the cylinder was an iron bar. When each revolution was completed the bar struck a bell. As each revolution meant that all the prayers within and without the wheel had been placed to the credit of the worshipper, the latter could tell at the end of each day, exactly how he stood to the credit of his account with the god (or shall I say, firm, they are so many), whom he believes to be responsible for his welfare and for his future state.

When we returned to the hall of audience, Mrs. Hari Chand, the "queen" of Lahoul, was ushered in, followed by a retinue of tag, rag, and bobtail, and her two little boys of four and two years of age. After much mutual *salāming*, another empty wine case was set up for her use, I had expressed a wish to see her jewellery, so with her came a round brass box about the size of a footstool, and which was set down in front of the Thakur. It was fastened with a padlock the key of which D. did not fail to remark was in Mr. Hari Chand's possession. He proceeded to unlock it and to display its contents, commencing by explaining that, in his opinion, and in that of all educated natives, the manner of wearing jewellery in vogue was absurd. He urged as an argument against the custom that if a native lady were wearing the nose and ear appendages, and her husband got drunk, he might do her serious harm by pulling them off, in which I cordially agreed with him. But the Thakur seemed to think that the piercing of the nose and ears for rings was dying out, and that in less than two hundred years would cease to be the fashion.

We had heard so much of the jewels worn by the queen of Lahoul, that we were much disappointed when we saw them. They consisted chiefly of nose, ear, and head ornaments. The necklaces were a long string of coral beads, interspersed with turquoise nearly black with flaws, and half a dozen pearls at intervals. There was a string of irregular pearls, and a bead necklet with an amulet attached. As I examined the coral and turquoise necklace, I inquired of the Thakur how it was that in a country where nearly all the people affected turquoises, one so seldom came across a really good stone. "Ah!" he replied, "the only valuable turquoises come from Iran (Persia), and how are we to get them?" He also told me that the coral came from Italy, where it was found growing under the sea, and was gathered by divers. I suggested that in my country the ladies did not care for the deep red coral. "Oh yes," he replied, instantly; "I know that the correct colour is a pale pink, or rose colour; but not being able to procure that, I was obliged to put up with the red." These remarks show how well-informed a man the Thakur is, for a native of the Central Himalayas.

The gems of the whole collection we considered to be the pearl necklace, and the gold amulet. The pearls, though irregular, were of good colour, large size, and alternately round and pear-shaped, but angularly shaped, not perfect. It took my fancy immensely, and I said this to the Thakur so often that D. trembled lest he should present it to me. In which case, in native fashion, we should have had to give him a *quid pro quo!* D. preferred the amulet box of gold filigree work, studded with turquoises, and with a ruby in the middle, which the Thakur told us came from Lhassa.

Meanwhile, as we were admiring the jewels, their owner, or perhaps I should say the owner for the time being (for the Thakur has the power, which he frequently exercises, of putting away one, and taking to himself another wife), sat looking on. The reigning queen was a small, good-looking child of perhaps sixteen, with regular features, good eyes, and teeth, and beautiful hands and feet, on which she wore the usual bell-anklets. She was dressed in red with a yellow *saree* on her head, embroidered in silk with flowers, which looked as if it had been made by machinery in England. Though she understood Hindustani, we could not get her to talk, and when D. suggested to her that she was a very lucky young woman to possess all this finery, she smiled a sad smile, as much as to say, "Ah, if you knew on how slender a thread my fortunes hang!"

She was not wearing much jewellery; but at the request of the Thakur removed the *saree* from her head to show us a very handsome and massive gold *kirkitzee*, filigreed within, and studded with precious stones. As we were examining it, she hazarded the remark that it was *boja* (heavy), as indeed I found, when I lifted it. A *kirkitzee* is a small saucer-shaped ornament of silver or gold, its weight and value corresponding to the wealth of the wearer. It is worn on the top of the head, by married women only, and corresponds to the wedding-ring of Christian countries.

When the jewels were again shut up in their brass box, and the queen had departed, accompanied by her retinue, we rose to take our leave. Hari Chand accompanied us downstairs, and, knowing him to be a travelled man, who had even trusted his precious person to the mercy of the sea, D. asked him to relate to us his experiences on that memorable occasion. He informed us that he had actually once been persuaded to take ship from Karachi to Bombay, a distance of five hundred miles, and that he was glad he had done it, for the sea should never carry him again. He went on to explain that he was armed with a letter of introduction from Lyall Sahib (Sir James Lyall), Lieutenant-Governor of the Punjab, and seemed surprised that, being possessed of so precious a document, the sea should have dared to molest him. We could imagine him waving the letter at the troubled waters, and expecting them instantly to be calm. Be that as it may, we gathered from him that it was unusually rough, and he further demonstrated, by word and gesture, that he was hideously sea-sick for two days and two nights: and this, in spite of the lieutenant-governor's letter in his pocket, and of some *sharab* the *sahib-log* on board had pressed upon him as a remedy. He told us that the captain had provided him with a small room, with a hole in the wall, to lie down in, but even that had no effect.

He began to look upon the missive of the governor-*sahib* as a fraud. But, at length, the journey came to an end, and the moment arrived to pay the fare. He went to the captain, and inquired how much he was indebted to him for so much pleasure. The latter replied that the usual fare was eighty rupees, but that from Hari Chand he would only take twenty-five. Radiant once more, the Thakur decided that the official letter was answerable for this reduction; but for which, I verily believe that the lieutenant-governor of the Punjab would have fallen seventy-five per cent, in the estimation of the king of Lahoul; and who knows what the effect might not have been upon the future of this country!

CHAPTER XXXII

OUR next march was from Kolang to Darcha, a short one of only eight miles. The road is excellent all along, and nearly level with the river. Our ponies walked it in two hours, very creditable for little beasts of thirteen hands. At the sixth mile the camping-ground lay but three hundred yards ahead. To reach it, however, we had to make a *detour* of two miles to reach a bridge. This was most annoying, especially as the rain was coming down in sheets. The camping-ground, or *parao*, as it is called, stands on a large sandy plain, at the south end of which meet the three rivers, the Bagha, the Kado Tokpo, and the Milang. Consequently, it is exposed to the winds which blow down these three valleys. In fine weather the dust beats anything I have ever seen. It was so fine that it was almost like smoke, and tin boxes, supposed to be air-tight, were filled with it; while, probably owing to the dust, all the four watches we had with us stopped during the two days we camped at Darcha. In bad weather we fared no better. From the exposed position of the camp, the rain came in every direction, and it was hard to know which way to face the tent.

There was a *serai*, at one end of which a small room eight feet square had been partitioned off for *sahib-log*. We rejoiced at first at the sight of that *serai*, and rushed for it when the foul weather came on. There was neither door nor window, but we tried to stop up the apertures with our tent *sulleetahs*. After half an hour's rain, however, the roof began to leak like a sieve, and we returned to our tents, which, if chilly at an altitude of twelve thousand feet, were, at least, dry. There was some good *chikor* shooting at Darcha. We found several coveys along the hillside above the *serai*, beyond the cultivation, and on the left bank of the Bagha, which we crossed by the most ricketty *jhula*-bridge I have ever negotiated.

From Darcha onwards the country assumed an entirely new aspect. When we left the last village behind, the pencil cedars became thinner and thinner, and, three miles further, ceased altogether. Eight miles beyond, the valley opened out into the grassy plain of Patseo. Here a fair is annually held among the traders from Rupshu, Ladakh, and far-away Rudok, on the one hand, and those of Kullu, Lahoul, and Spiti, on the other. The articles of commerce are chiefly wool, salt, corn, and borax, and much business is done. During the two months which the fair lasts, it is supposed to be visited by the assistant-

The Last Rest-House in Lahoul

commissioner in Kullu, whose business it is, on behalf of the British Government, to see that no unlawful taxes are levied on the traders, and that these poor ignorant creatures are not unduly fleeced by the powerful and better-educated Thakurs of Lahoul, in whose country the fair takes place.

Beyond Patseo the road took a direct turn to the east, ascending rapidly for four miles to the halting-place at Zingzingbar, at the foot of the Bara-Lacha Pass, and fourteen thousand two hundred above the level of the sea. It had been selected for the camping-ground, on account of the fresh-water spring there; but there was no level ground about, except a few platforms which had been scooped out of the hillside to pitch tents on. A more uninviting-looking place than Zingzingbar cannot well be imagined. A waste of rock and stones, the ground was too hard for the tent-pegs to be driven in, and the ropes had to be made fast to the rocks and boulders. Across the river, opposite the camp, there was a pretence at grazing ground, and through the stream, accordingly, *nolens volens*, all the baggage animals were driven. They were washed away fifty or sixty yards down stream ere they could gain a footing on the opposite bank; but a shake seemed to restore their equanimity again, and without more ado they started off in search of a few blades of grass.

Several wild-looking travellers passed our camp. One, in particular, attracted my attention by the beautiful turquoise earrings he wore, the colour of which I noticed, though unable to inspect them closely. I sent a servant after him to offer to buy them; but, as usual, he declined to sell, as they belonged to the wife he had left behind him. It is the custom for these traders when making far-off journeys to wear the jewels belonging to their wives, or which are known in their own country as heir-looms. They will never part with them, and should they die on the road, the jewels are brought home by their comrades as proof of their decease.

We camped for one night at Zingzingbar, and on the 27th of August continued our march to the summit of the pass, two thousand three hundred feet above us. The ascent was very gradual; and, though the Bara-Lacha is sixteen thousand five hundred feet above the sea, it is the easiest pass I have ever gone over. The distance from Zingzingbar was some seven miles. The road begins with three or four zigzags for some five hundred feet or so, and thence rises gradually to the summit, passing immediately under two or three glaciers. These, fortunately, had not to be crossed; but a succession of streams issuing from them gave us much trouble, for they have no fixed beds, and cannot be bridged. Three miles further on we crossed the Bagha, which, though here unfordable, is a comparatively insignificant stream, by a wooden bridge. Then, after passing up a narrow gorge with a natural dam at the head, we came suddenly upon a lovely lake of deep blue water, many acres in extent, and most appropriately named the Suraj Dal, or Sun Lake. It is fed at the upper or eastern extremity by a glacier stream, and, as at first appeared, had no visible outlet. A closer examination, however, showed us the waters escaping by a sub terranean duct into the narrow gorge up which we had come. The Sun Lake is sixteen thousand and sixty feet above the sea; and, though the surrounding mountains are absolutely devoid of trees or vegetation of any kind, the reflection in its deep blue waters of the snowy peaks that encircle it made a beautiful picture, that can never be effaced from our memories. While we halted on its banks, D. managed to take a photograph, while I amused myself with gathering edelweiss, of which there was abundance.

A mile further on we reached the summit of the pass. It is marked by a large cairn of stone, engraved with the ever-recurring prayer, "Om mani padmé hum," and covered with dilapidated prayer-flags, held down with rocks, as

"Suraj Dal" or Sun Lake, at top of Bara-Lacha Pass, 16,060 Feet

there is no wood within many a long mile to serve as a flag staff. Clouds were banking up, and the cold was so searching that we were glad to crouch in the shelter of the shrine while we enjoyed the magnificent view. Great glaciers to right of us, glaciers to left of us, glaciers in front of us, and behind the Sun Lake.

Two high-roads (?) here diverge, one leading north to Ladakh, and the other south to Spiti, and used only by the Spiti traders. The other, however, is the medium of a very considerable traffic, not only local, but between Amritsur, the Sikh capital of the Punjab, and Leh, the capital of Ladakh. The double journey occupies from three to four months, and considerable hardships are endured by the travellers. The most deadly enemy the traders have to contend with is a sort of typho-pneumonia, and which is so suddenly fatal, and so contagious, inspiring such a panic, that we were informed on the highest authority that it was not unusual for those attacked by it to be buried alive, so as to get them out of the way as soon as possible! For buried alive, however, read, laid on the ground, and covered with stones, so that if not actually dead the poor wretches endure intense and prolonged agony. The regular traders on this road know the places where living burials have taken place, and nothing will ever induce them to approach them; for they say that for years after such a sacrifice has been made, shrieks can be heard proceeding from the spot.

Though the Spiti road is remarkable for its magnificent scenery, the route is not a favourite one with European sportsmen, on account of the number of large and dangerous glacier streams which have to be forded. Natives are often carried away and lost; but they seem to attach less value to the life than to the clothes and jewels. Egerton, in his book, tells a good story of an adventure which happened to him when crossing one of these streams. A couple of coolie women, carrying his baggage, were swept away down the stream. Egerton ran down the bank, and at considerable danger to himself, rushed into the water and succeeded in saving one of them. When he returned by the same road a few weeks later, the woman he had rescued came to see him. At first he thought it was to thank him for having saved her life, and was pleased at the exhibition of gratitude. But no, not one word of thanks did he get, only a demand for compensation for loss of ornaments and clothing! From this true anecdote a very fair estimate can be obtained of the character of the natives of India, and I commend it to those, and there are many of them, who stand up for them without any personal experience of them.

The snow now began to fall close to us, and we thought we had better retrace our steps. We had not descended more than a mile before we heard shouts behind us, and, looking round, saw young Mr. P—, who had passed through Kailang just two months before on his way to Changchenmo. He had succeeded in bagging three Thibetan antelopes and a small *burrhel*, and looked as fit as a fiddle, in spite of all the skin being burned off his face by the sun and the wind. He accompanied us back to Zingzingbar, and stayed three or four days with us there and at Darcha.

On the way down we passed a small hut, in which Captain M—, when crossing the pass earlier in the season, found a human skeleton. But there was nothing in it as we went by. Half way down it began to rain and sleet and grew horribly cold; but we soon reached the shelter of our tents and laughed at the elements over a pot of hot soup.

The weather grew worse. All night it poured, and towards morning it was snowing. On such a day it would have been madness to march, except in a case of sheer necessity, such as was ours. We had neither firewood to cook with, nor forage for our ponies. Both of these commodities we had been compelled to bring with us, as we had left the last village of the country behind us, and before us lay a desert. We had estimated our requirements at four days' supply only. The four days had now elapsed, and there was nothing for it but to return. The firewood difficulty we might, indeed, have overcome by burning our tent-pegs. But we had made all our available flour into "chuppatties" for the ponies, and there was now no food left for them, so we packed our kit and started. The cold was so intense that we could not even feel the reins, and there was no alternative but a walk of fourteen miles. As we descended the snow turned into sleet, the sleet into rain, and such rain, and such a gale! We were soon soaked to the skin. The longest lane, however, has a turning, and in due course we reached the village Darcha, where we lost no time in lighting a fire and endeavouring to dry ourselves. Many hours afterwards the baggage turned up and we were able to get at a change of clothes. We pitched the tents as quickly as possible and speedily forgot our troubles.

As regards the commissariat, however, we found ourselves in rather a predicament. We had just finished all our meat, and though we had another sheep it was not killed, which was a trifle awkward, especially as we had to feed Mr. P—. I mentioned the difficulty to him, and he promptly volunteered to tell his servant to make over to ours the two or three joints of mutton which he had with him, for our common table. The following morning, however, my *khansamah* came to me with a long face, and told me that the whole of the meat had been eaten by dogs in the night. I simply did not believe the story, and, indeed, afterwards ascertained that instead of handing the joints over to my cook, as ordered, Mr. P—'s servant had given a dinner-party with it the preceding night. In consequence we were left entirely without meat, while the wretches who had robbed us got off scot-free, for it was impossible to fix the blame on any particular individual. Such are natives!

Immediately in front of our camp at Darcha, on the right hand of the Milang river, stands an enormous mound of debris, supposed to cover the ruins of the old village of Darcha, destroyed a century ago by an avalanche. The glacier at the head of the Milang *nala* slipped, and an enormous mud avalanche came tearing down the valley, completely overwhelming the village, not a soul escaping. But this happened on the other side of the river, and the mound above alluded to must therefore be of more recent date, and was caused, probably, by a landslip of the whole mountain side, heralded, however, by the warning fall of incessant boulders, without any apparent cause.

We remained a few days at Darcha, replenished our robbed larder with *chikor*, and then returned to Kolang on the 31st of August. Here we enjoyed three days of the vilest weather imaginable—snow, rain, sleet. When the snow came on, D. sent up and begged the Thakur to hire him two men to sit day and night by our tents and sweep the snow off the roofs. But for this timely precaution we should infallibly have had our tent poles broken.

We were not a little surprised when the weather cleared, and we walked about the fields where the hay and wheat crops lay sodden on the ground, to find the peasants making no attempt to turn and dry them in the sunshine. D. asked the Thakur the reason of the waste. It seems attributable a great deal to fatalism. The ignorant native custom is to let crops, once soaked, lie and rot on the ground! Lahoul used to be a remarkably dry country, as I have before mentioned. To within the last few years the average rainfall was but a quarter of an inch annually. Of recent years it has much increased, and in 1892 must have been several inches. Rain, however, is rare in harvest-time. Not so, snow;

snow lays the crops level with the ground, and there is nothing for it but to leave them till the snow melts. In their ignorance, I presume, they do not distinguish between the effects of rain and snow, and pursue the same course in both cases.

CHAPTER XXXIII

WE returned to Kailang on the 12th of September, and having, as previously described, accepted the invitation of the Thakur Hari Chand to visit him at his castle of Kolang, we found ourselves obliged to pay a visit to the Thakur of Gumrang, Devi Chand. The two "castles" were much alike in every detail except that that of Gumrang in its private chapel boasted a massive silver chorten ornamented with gold and precious stones. The chorten had been made at Lhassa, the "Mecca" of Thibet, whence it was imported into Lahoul by an ancestor of the Thakur's, who looked upon it with immense pride. Devi Chand has recently married a seventh or eighth wife, and become the father of an only child, a pretty little girl, to whom he was devoted. In consequence, he had attempted, unsuccessfully, to break through an immemorial Lahoul law, which provides that a daughter cannot inherit landed property. Among the zemindars, or farmers, who die without male issue, the land lapses to the Thakur of the jaghir in which the former lives. Being himself perniciously affected by a law by which he had often previously benefited, Devi Chand sought to set it aside. Not succeeding, he was compelled to adopt as his heir the son of a brother with whom he had always been at deadly enmity. As the result of perpetual litigation, in which the Thakur was eventually successful, the brother was banished to a small house near Darcha, where he soon afterwards hung himself. So long as there was a chance of Devi Chand's machinations on his daughter's behalf proving successful, he lost no opportunity of abusing his nephew, even giving out that he was not his brother's son, but that of his brother's wife and his brother's groom. When he lost his case, however, the Thakur made the best of a bad job and now lives amicably with his heir.

He received us very hospitably, and in addition to the usual offering of dried fruit, pressed upon us two country-made blankets, which we gave away directly we got back to camp. We had to return his generosity, however, by the present of a couple of Roman shawls! The Thakur was very anxious that D. should take a "sun-picture" of his little girl, but, unfortunately the sun would not oblige.

We left Kailang in September, and started on our twenty days' march for the nearest railway station, *en route* for old England once more. We met on the

Gumrang Castle, the Residence of Thakur Devi Chand

road a striking example of the hold which the sacerdotal class has on the people of this country, in shape of a *lama* riding along, as drunk as a lord. Yet to this creature, one after another, the poor coolies, uncovered as they passed, begging for his blessing on bended knee.

Truly the ways of the Lahouli agriculturist are wonderful! We noticed that instead of reaping the corn, they pulled it up by the roots, and shaking off the earth, laid it flat on the ground in such a way that the ears are underneath. There they leave it for a couple of months before they gather it in. The roots the cattle gradually devoured for fodder, and in a dry season the method may answer. But in this abnormal year, with a rainfall of two hundred per cent, above the average, the loss must have been enormous, and in any case the process must impoverish the land.

The outlook was indeed bad. All through September it had rained almost incessantly. The mountains long since had been covered with snow, and the cold weather had set in. Yet not a third of the hay crop was cut. This meant starvation to hundreds of poor animals, and, in addition, a scourge of rinderpest. The treatment of this disease appeared to be the boiling down of the flesh of cows and calves which had died of it, and the administering the meat to those not affected. On inquiry, we failed to ascertain that this remedy was efficacious!

Before I left Kailang I had made for me by the local jeweller, who was also the blacksmith, a *kirkitzee*, or Lahouli married women's head ornament. I gave the workman the silver in rupees, and when the ornament was finished he weighed it in my presence, to show that it contained the equivalent in silver, and I paid him for his workmanship two *annas* in the rupee. Now in Bombay they demand one rupee for work on each rupee's weight of silver, and the latter is not pure silver, as in Lahoul.

We left Koksur on the Lahoul side of the Rohtung Pass on a sharp frosty morning. It was snowing slightly, but we crossed the pass without any difficulty, though a few days previously some natives attempting it had been compelled to beat a retreat, driven back by the fierce wind.

For five months we had sojourned in a virtually treeless country, and the view from the top of the pass, looking towards Kullu, with its dense forests, its smiling fields and undulating pastures was indeed refreshing. Perhaps the most striking feature in the landscape was the brilliant colouring of the countless fields of red millet. This beautiful plant, *suryari*, grows to a height of six or eight feet, and the flower or fruit, of a brilliant scarlet colour, and eighteen inches long, hangs in graceful festoons from the stem.

We reached Rahla at the foot of the pass shortly after noon, and took up our quarters in the resthouse, revelling in the unaccustomed luxury of dining off china plates, after such a long use of iron enamel.

We had left Lahoul, and much regretted that necessity had compelled us to do so, for surely there are few countries on this much-favoured earth that can boast of a climate so healthful and so invigorating, or of scenery so grand, so varied, and, to lovers of the wild, so enchanting, as that we had now quitted, perhaps never to visit again. We had left, too, with much regret our friends at the mission house at Kailang, but we shall carry them with us, in our thoughts, at least, for many years to come.

INDEX

MAP

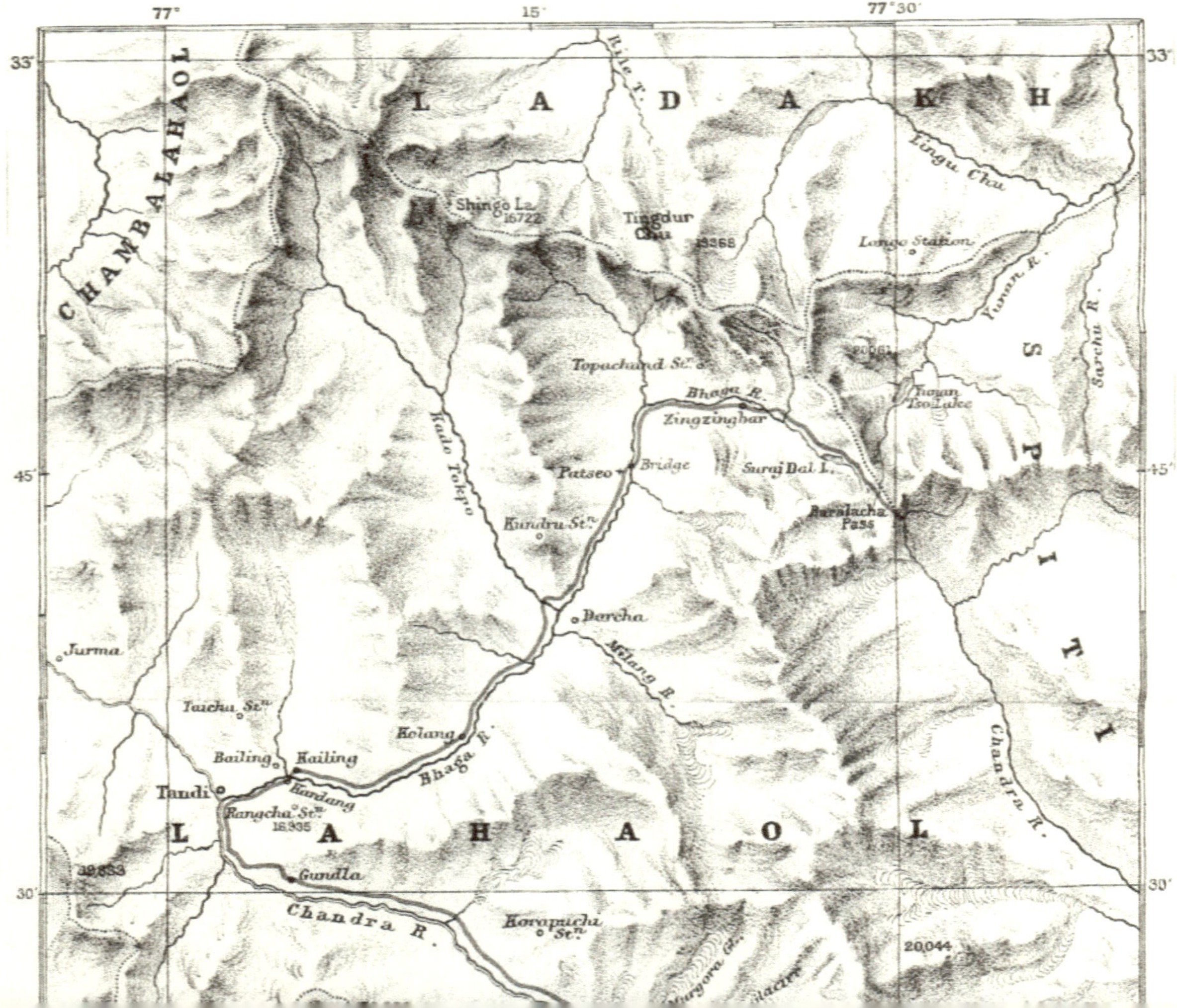

77°
15'
77°30'
33'
45'
30'
LADAKH
SPITI
LAHAOL
CHAMBA LAHAOL
Rile T.
Lingu Chu
Lango Station
Yunan R.
Sarchu R.
Shingo La
16722
Tingdur Chu
18368
20061
Yunan Tso Lake
Topachand Stn.
Bhaga R.
Zingzingbar
Bridge
Surajdal L.
Patseo
Baralacha Pass
Kundru Stn.
Kado Tokpo
Darcha
Milang R.
Chandra R.
Jurma
Taichu Stn.
Kolang
Bailing
Kailing
Tandi
Kardang
Rangcha Stn.
16,335
Gundla
19,833
Koropuchu Stn.
20,044

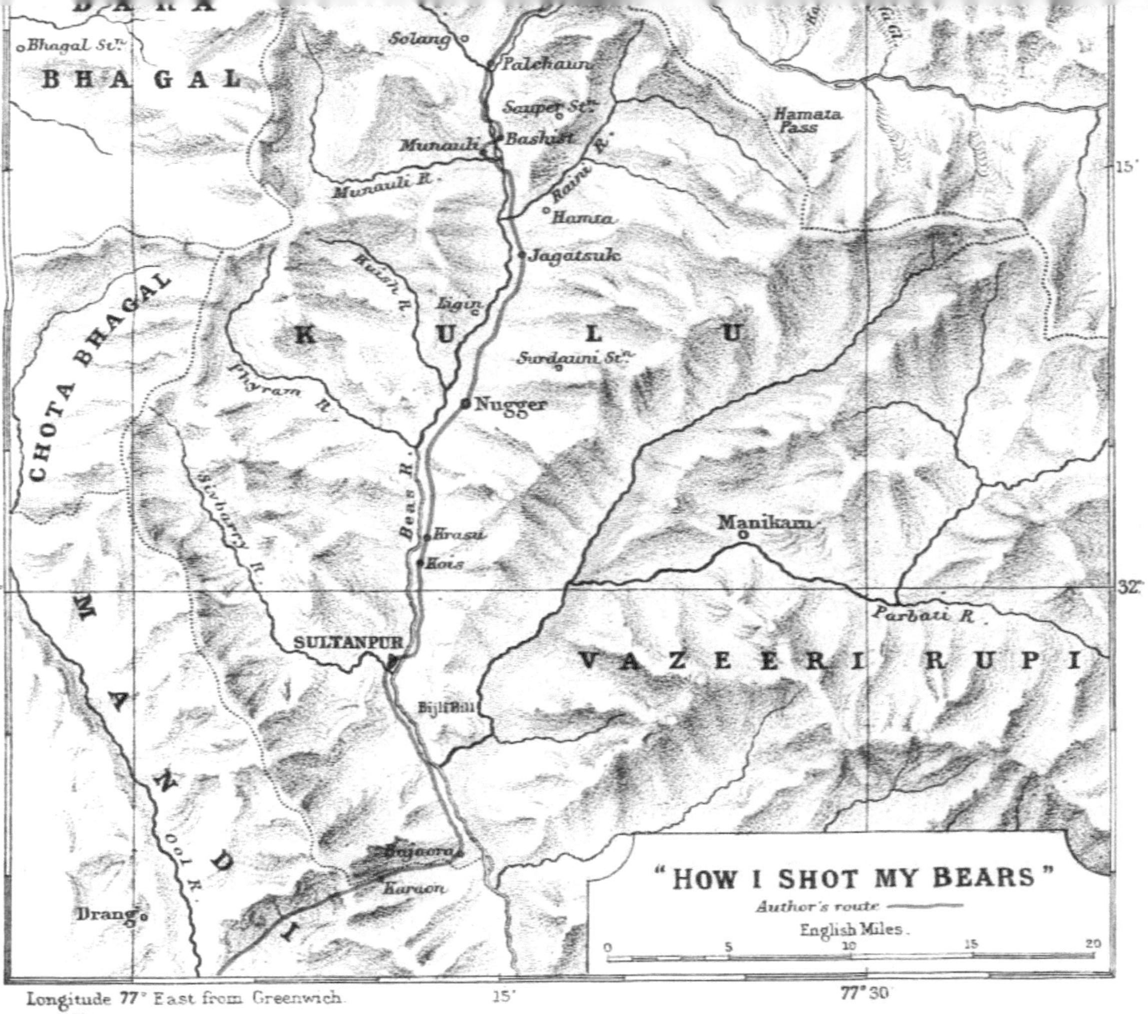
"HOW I SHOT MY BEARS"
Author's route
English Miles.
0 5 10 15 20
BHAGAL
Bhagal St.n
CHOTA BHAGAL
KULU
VAZEERI RUPI
MANDI
Solang
Palchaun
Sauper St.n
Bashist
Munauli
Munauli R.
Raini R.
Hamta
Hamata Pass
Jagatsuk
Kuish R.
Ligin
Surdaini St.n
Nugger
Phyran R.
Beas R.
Krasu
Kois
Manikam
Parbati R.
Sivbarry R.
SULTANPUR
Bajaora
Karaon
Ool R.
Drang
15'
32°
77° 30'
Longitude 77° East from Greenwich

Made in United States
Troutdale, OR
11/03/2026

56844874R00132